Flow Noise

Con Doolan · Danielle Moreau

Flow Noise

Theory

Springer

Con Doolan
The University of New South Wales
Sydney, NSW, Australia

Danielle Moreau
The University of New South Wales
Sydney, NSW, Australia

ISBN 978-981-19-2486-6 ISBN 978-981-19-2484-2 (eBook)
https://doi.org/10.1007/978-981-19-2484-2

This Springer imprint is published by the registered company Springer Nature Singapore Pte Ltd.
The registered company address is: 152 Beach Road, #21-01/04 Gateway East, Singapore 189721, Singapore

To those who have inspired us and to those who will read this book.

Preface

This book was originally conceived as a textbook to assist new postgraduate students joining our research group. Many engineering graduates are attracted to aeroacoustics and hydroacoustics (flow noise) problems and research. However, many of our students come from diverse educational backgrounds that may not cover the necessary background material. Also, we found that many undergraduates are very keen to work on flow noise research projects in their final year, but need additional support to learn the fundamentals of the subject. This is because flow noise is a combination of unsteady fluid mechanics and acoustics, both of which are normally part of elective streams at undergraduate level and it is rare for students to specialise in both by graduation.

As we wrote this book, it became clear that there are many practicing engineers and researchers that need to work in flow noise problems but do not have the required fundamental knowledge, either through not being taught it originally or that they require a refresher as they have been working in other areas.

Thus, this book aims to provide an introduction to aeroacoustics and hydroacoustics theory (we call this flow noise). It is a little different from other books in the discipline as it assumes the reader has undergraduate-level mathematics skills and only a small amount of fluid mechanics background. It takes the reader through a lot of fundamental material first before developing the major theoretical concepts in flow noise. We have attempted to make the book as accessible as possible so that students and professionals can learn the required background in flow noise theory. It is hoped that this can be used as a foundation for their work and for further study in some of the more advanced textbooks.

Writing this book has not been possible without the students, research associates (postdocs) and other staff associated with the Flow Noise Group at the University of New South Wales, Sydney. Special thanks must be given to Omear Saeed, our laboratory Technical Officer, who keeps everything running in the lab so smoothly it has allowed us to write this book. There are almost too many past and present students and colleagues to thank: Ahmed Mahgoub, Allan Harrland, Angus Wills, Branko Zajamsek, Chaoyang Jiang, Charitha de Silva, Chung-Hao Ma, Cristobal Albarracin, Damien Leclercq, Elias Arcondoulis, Jesse Coombs, Jeoffrey Fischer, Jiawei Tan,

Karn Schumacher, Laura Brooks, Manuj Awasthi, Marcus Boyd, Mohamed Sukri Mat Ali, Prateek Bahl, Roman Kisler, Rowena Dixon, Zebb Prime, Ric Porteous, Sean McCreton, Shakeel Ahmed, Tingyi Zhang, Warrick Miller, Yendrew Yauwenas, Yuchen Ding and Ziao Zhang.

Finally, we must thank the generous financial and technical support of the agencies and companies that have funded our work over the years. It is this support that has enabled so much research and teaching in flow noise to be possible. We would especially like the thank the Australian Research Council and the Defence Science and Technology Group for their support.

Sydney, Australia
March 2022

Con Doolan
Danielle Moreau

Contents

Chapter 1
Introduction

Abstract This chapter provides a brief introduction of the textbook. The introduction chapter is in two sections. The first section explains what flow noise is and why it is important. The second section gives an overview of what is contained in each chapter of the book.

1.1 Flow Noise

Flow noise is simply the sound created by fluid flow. It is also known as aeroacoustics (the sound created by airflow) or hydroacoustics (the sound created by water flow). Here, we take the opinion that the underlying physics is the same regardless of the medium, hence we use the term flow noise.

Flow noise is a unique branch of science and engineering that draws heavily on both fluid mechanics and acoustics. As sound is created by unsteady flow processes (in most cases), someone working in the area requires a good understanding of both steady and unsteady fluid mechanics. Further, a background in acoustics is required. Bringing these disciplines together requires a good foundation in vector calculus and spectral analysis. These tools are used to analyse laminar and turbulent flow and link them to sound production. This book aims to provide the reader with a solid basis in the background material before presenting the core theory in flow noise.

Perhaps the best known example of sound created by an unsteady fluid flow is jet noise; the intense noise created by jet engines used for civilian and military aircraft. Jet noise is a wonderful example because it is sound created by flow itself with no interaction with solid boundaries. It is often assumed by many acousticians and fluid mechanicists alike that flow noise must involve a vibrating boundary driven by flow oscillations. This is not the case. As we will explain, unsteady fluid flow creates quadrupole sources of sound, formed by the fluctuating fluid velocity field, with no need for any solid surface.

Once a solid rigid surface is introduced into a flow, the opportunities for noise production increase. A simple example is a rigid cylinder placed in cross-flow. This seemingly simple example is actually quite complex and is an excellent case to explain many aspects of flow noise. A cylinder will disturb a uniform flow and create

C. Doolan and D. Moreau, *Flow Noise*,
https://doi.org/10.1007/978-981-19-2484-2_1

an unsteady wake. This unsteady wake induces an unsteady force on the cylinder. We will show that unsteady pressure and forces on a surface support dipole sources of sound. These are much more efficient than the quadrupoles mentioned above. This means that significant levels of sound can be created by solid objects in low Mach number flows that we would not normally associate with high levels of noise production. If the solid object was not available to support these more efficient dipoles, it is likely that the level of noise generated would be negligible. Note that we do not require the cylinder to vibrate or move to support these dipole sound sources.

The cylinder example shows us what is important in understanding flow noise. The unsteady wake is quite often turbulent, thus an excellent understanding of turbulence is needed. This includes an understanding of the turbulent statistics, correlations and spectral representation of the turbulent flow field. Once these are known, noise can be estimated by relating these flow quantities to the dipole source strength. Acoustic theory is then used to estimate the acoustic field radiated by the cylinder. Cylinder flows can be used as a proxy for understanding practical examples such as aircraft landing gear, cables used on bridges, automobile external mirrors and protuberances from marine vessels.

Another important example of flow noise is the sound created by the interaction of turbulence with an edge. A very good example of this is the phenomenon of airfoil trailing edge noise. Here, as turbulence moves past an edge, it experiences a sudden impedance change. Diffraction of the near field pressure field ensues and results in the radiation of sound. In some ways, the edge can be thought to act like an old-fashioned record needle. The needle senses the unsteady surface of the record groove which is turned into sound using an amplifier. In the case of edge noise, the edge 'feels' the unsteady flow of nearby turbulence (like a needle) and the diffraction mechanism converts this unsteadiness into sound (like an amplifier). So, when we hear edge noise from a turbulent flow, we are hearing turbulence itself. This is nearly always a hissing broadband noise, proving to us once again that turbulence is random.

You may hear the random, broadband noise associated with airfoil trailing edge noise in everyday life. If you pass close enough to a wind turbine, you will notice a 'swooshing' hissing sound. This is the turbulence in the blade boundary layer turned into sound. It has the 'swooshing' characteristic due to the additional effect of relative motion between the source (blade edge) and the observer. This effect includes a Doppler shift in frequency, an increase in level due to convective amplification and the result of the unique directivity of trailing edge noise. Other examples of airfoil trailing edge noise can be found in aeroengine fan blades, propellers (helicopter, drone and marine) and cooling fans used in computers, air-conditioning systems, buildings and automobiles.

Edge noise may also be created when unsteady flow interacts with the leading edge of an airfoil or in fact any solid object. In the case of turbulence interacting with an airfoil, we call this airfoil leading edge interaction noise, or turbulence-interaction noise. Normally this type of noise is a combination of surface dipoles created by the distortion of turbulence interacting with the edge as well as diffraction of small-scale turbulence. This occurs in many applications, particularly propellers and fans placed in turbulent wakes or within ducts. Wake interaction with a moving surface, such as a propeller blade is particularly important. Wakes generated by rotors may interact with downstream stators (or vice-versa), creating noise with a tonal quality. This is due to the unsteady forces induced on the blade passing through the wake velocity deficit created by the upstream object. A related noise mechanism may occur in tandem aligned cylinders (aircraft landing gear) or any situation where one object is placed in the unsteady wake of another.

When a solid object moves with respect to the observer, such as in the case of a propeller, more important sound sources are produced. So called ‘thickness noise’ is created by the relative displacement of the blade with respect to the observer. As this is a volume displacement, this is a particularly efficient radiator of sound, known as a monopole. Thickness noise is proportional to the rate of volume displacement, so is more noticeable on faster rotors like helicopter main rotors than slow moving devices, such as wind turbines.

A different type of noise is created by the relative motion of a steady force and an observer. This situation typically occurs in the case of a propeller. A propeller develops steady thrust by creating steady forces on each blade; however, these steady forces rotate about a fixed axis and may move relative to a nearby observer. This relative back-and-forth movement of these forces creates ‘steady loading noise’, normally a tone and harmonics associated with the rotational speed of the propeller. Steady loading noise is a feature of aeronautical propellers and is particularly prominent in drone propeller noise signatures. Steady loading noise is a dipole source.

Flow noise sources may be placed with an enclosed volume. A common type of enclosure is a circular or rectangular duct. A duct changes how the acoustic source couples with the acoustic medium about it. Further, the amount of noise that escapes the duct exits is dependent on the type of source, its frequency and the dimensions of the duct. Ducts are found in many aerospace applications; however, ducted marine propellers are also becoming increasingly common.

The short discussion above presents the main flow noise mechanisms that are described in this book. They are associated with many practical applications and a thorough understanding of their underlying physics is needed in order to design and control low noise flow devices. These mechanisms are, however, associated with unsteady flow and its interaction with rigid surfaces. There are a few additional noise sources associated with the coupling of unsteady flow with vibrating surfaces. While very important, these are outside the scope of the present book. Further, the creation of sound by bubbly flows is also important, but the authors decided to focus on the core aspects of flow noise for an introductory textbook.

1.2 Overview of This Book

This is an introductory textbook in flow noise theory. The book concentrates on foundation material as much as flow noise. The book has been designed this way to make flow noise more accessible to engineers and scientists who may not have had these topics included in their training, or are in need of a refresher.

Vector calculus is covered in Chap. 2, which is a concise presentation of fields, tensors, vector operations and some important integrals. Chap. 3 introduces the reader to the fundamentals of spectral analysis, the most important being the use of the Fourier transform.

Both fluid mechanics and acoustics are controlled by the fundamental equations of fluid mechanics and these are derived and shown in Chap. 4.

This brings us to the foundation chapter in acoustics, Chap. 5. This important chapter covers sound waves, decibel representation, the acoustic wave equation and its solution. It also introduces the reader to simple sound sources.

Laminar and turbulent flow is another foundation topic for flow noise and this is presented in Chap. 6. Some fundamental laminar flow fields are described. Transition and instabilities are covered before the chapter moves into the realm of turbulent flow. The classical description of turbulence in terms of an energy cascade over multiple scales is adopted. This is used to provide a basis for understanding the statistical quantities of turbulence, including correlations in space and time and spectral models. Reynolds averaging, turbulent boundary layers and turbulent wall-pressure spectral models are also included.

Chapter 7 builds on the introductory and foundation material to present the fundamentals of flow noise generation. After covering some more preliminary theory and defining compact multipole sources, noise from turbulent flow is presented in the form of Lighthill's famous theory. Noise from mass-injection and unsteady forces are then presented as general solutions to the wave equation. Curle's theory is covered which gives the reader the ability to understand and analyse the case of solid rigid objects placed within unsteady flow. An informative example is presented where the flow noise of a cylinder is analysed, linking the turbulent flow properties to noise generation. Finally, this chapter covers the important Ffowcs Williams and Hawkings equation, allowing the analysis of moving sources. The Ffowcs Williams and Hawkings equation adds steady loading and thickness noise sources to our repertoire.

Airfoil noise sources and control mechanisms are presented in Chap. 8. Airfoil leading edge and trailing edge noise are covered first, including laminar, blunt trailing edge and turbulent trailing edge noise sources. A comprehensive section reviewing noise control techniques is then provided.

The final chapter, Chap. 9, concerns duct acoustics. In particular, the fundamentals of duct acoustics are covered, followed by descriptions of what occurs when sources are placed inside ducts and how sound can be radiated away from duct exits. A simple scaling model is presented, which gives guidance on the effect of a duct on dipole sources.

Chapter 2
A Review of Vector Calculus

Abstract This chapter provides a review of vector calculus, used to derive the fundamental equations of fluid mechanics, acoustics and flow noise (aeroacoustics). It is intended introduce to the reader the mathematical building blocks for the rest of the book. The chapter begins by introducing scalar, vector and tensor fields. Tensor notation is revised followed by a summary of important vector operations. The use of the 'del' operator is described. The chapter finishes with details concerning line, surface and volume integrals and important theorems that relate these integrals to each other. The Dirac delta function is included.

2.1 Introduction

Vector calculus is an important part of engineering mathematics and is especially important for the fundamentals of flow-induced noise. This chapter, while only brief, summarises the important areas of vector calculus needed for the remainder of this book.

The chapter begins with the basic definitions of scalar, vector and tensor fields. At this point, tensor notation is introduced and defined. This is a widely used methodology for describing tensors and vector operations.

Basic vector operations are defined first, such as the dot and cross products and the scalar triple product with a description of the appropriate tensor notation for each.

After the basic vector operations, the chapter introduces the 'del' operator, the way we obtain partial derivatives of scalar, vector and tensor fields. These include the important gradient, divergence, curl and Laplacian operators. The Kronecker delta is also introduced at this point to illustrate its usefulness in vector operations.

Line surface and volume integrals are defined before the common theorems that are used to relate them to each other. Finally, a definition of the Dirac delta function (or the unit impulse function) is defined at the end of the chapter.

C. Doolan and D. Moreau, *Flow Noise*,
https://doi.org/10.1007/978-981-19-2484-2_2

2.2 Fields and Field Notation

Three types of *field* are required for the description of flow and acoustics. These are *scalar, vector and tensor* fields [1, 2].

2.2.1 Scalar Fields

Scalar fields describe the variation of scalar quantities in space and time. For example, a scalar potential field ϕ can be described by,

$$\phi = \phi(\mathbf{x}, t) = \phi(x, y, z, t) \tag{2.1}$$

where $\mathbf{x} = (x, y, z)$ is the vector representation of a point in a Cartesian coordinate system, and t is time.

2.2.2 Vector Fields

Vector field describe how vector quantities (those with both magnitude and direction) vary in space and time. An example is a force field $\mathbf{F}$,

$$\mathbf{F} = \mathbf{F}(\mathbf{x}, t) = \mathbf{F}(x, y, z, t) \tag{2.2}$$

Note that a vector field can also be described in terms of its individual components. For a Cartesian coordinate system,

$$\mathbf{F}(\mathbf{x}, t) = F_x(\mathbf{x}, t)\mathbf{i} + F_y(\mathbf{x}, t)\mathbf{j} + F_z(\mathbf{x}, t)\mathbf{k} \tag{2.3}$$

where $F_x(\mathbf{x}, t)$ is a scalar field that describes the $x-$component of the force vector at position $\mathbf{x}$ and time t. Similarly, $F_y(\mathbf{x}, t)$ and $F_z(\mathbf{x}, t)$ describe the $y-$ and $z-$ components at at position $\mathbf{x}$ and time t. The symbols $\mathbf{i}, \mathbf{j}, \mathbf{k}$ represent the unit vectors in the x, y, z directions, respectively.

2.2.3 Tensors and Tensor Notation

Tensors are useful constructs that allow us to generalise and manipulate scalar, vector and matrix quantities [2]. Tensors have *rank*, which is an integer number to describe the type of number system we are using. A scalar is a tensor of rank 0, and a vector is a tensor of rank 1. Tensors of higher rank can be formed, but we will normally use second rank tensors: 3×3 matrices in this book. We will drop the "second rank" and normally refer to these matrices as tensors.

Tensor notation is a compact way to write and manipulate vectors and tensors. It uses the rank of the tensor as a subscript to efficiently describe the vector quantity. A scalar is a rank 0 tensor, thus it has no (zero) subscripts. An example is the potential field ϕ described in Eq. (2.1).

A vector is a rank 1 tensor and has one subscript. For example, in three-dimensional space each position will have three coordinates and is represented by vector $\mathbf{x} = (x, y, z)$. In tensor notation, this is written,

$$\mathbf{x} = x_i \tag{2.4}$$

where the subscript $i = 1, 2, 3$ defines each component of the vector. For example, in a three-dimensional Cartesian coordinate system, $x_1 = x$, $x_2 = y$ and $x_3 = z$.

A tensor with rank 2 has two subscripts. This a tensor may be written as Q_{ij}; where i refers to the row and j refers to the column (a matrix). As an illustration of how to use a tensor in a three-dimensional Cartesian coordinate system, say we wish to describe the stress field inside a material. The Cauchy stress tensor [2, 5] is used for this purpose,

$$\boldsymbol{\sigma} = \sigma_{ij} = \begin{bmatrix} \sigma_{11} & \sigma_{12} & \sigma_{13} \\ \sigma_{21} & \sigma_{22} & \sigma_{23} \\ \sigma_{31} & \sigma_{32} & \sigma_{33} \end{bmatrix} = \begin{bmatrix} \sigma_{xx} & \sigma_{xy} & \sigma_{xz} \\ \sigma_{yx} & \sigma_{yy} & \sigma_{yz} \\ \sigma_{zx} & \sigma_{zy} & \sigma_{zz} \end{bmatrix} = \begin{bmatrix} \sigma_{xx} & \tau_{xy} & \tau_{xz} \\ \tau_{yx} & \sigma_{yy} & \tau_{yz} \\ \tau_{zx} & \tau_{zy} & \sigma_{zz} \end{bmatrix} \tag{2.5}$$

The nine components of the Cauchy stress tensor in Eq. (2.5) completely describe the stress field at a point inside a material. Figure 2.1 shows how the stress terms

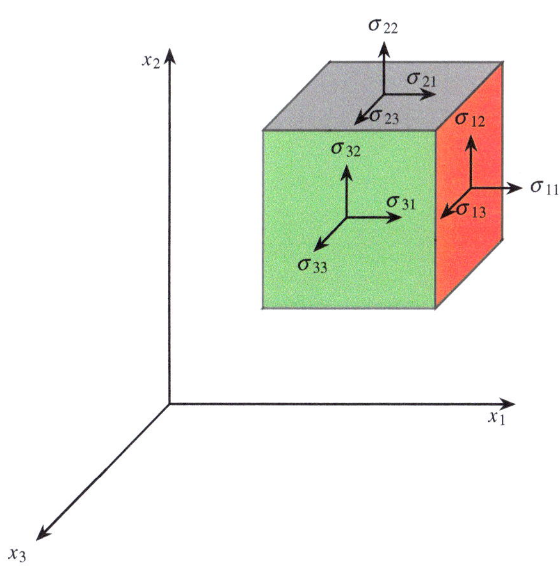

Fig. 2.1 The components of a stress tensor in tensor notation

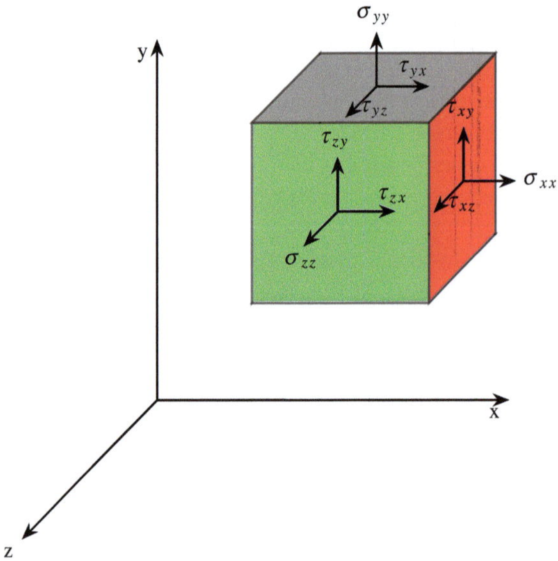

Fig. 2.2 The components of stress that comprise the Cauchy stress tensor using material stress notation

are labelled using tensor notation. In Fig. 2.2, we have replaced σ with τ for the off-diagonal terms of the tensor in order to remain consistent with naming conventions for stress fields in materials. This illustrates how we use the subscripts to uniquely define each element of the tensor. For example, $\sigma_{11} = \sigma_{xx}$, $\sigma_{32} = \tau_{zy}$, and so on.

While we conclude this section on tensor notation at this point, we will revisit and expand upon tensor notation in this chapter and throughout the book. This is because tensor notation is a very powerful (and simple) way to describe vector operations.

2.3 Basic Vector Operations

This section provides a concise overview of common vector operations [1, 4] that are used in this book. We will use common and tensor notation, as appropriate, in the representation of vectors and scalers in these operations.

2.3.1 Vector Addition and Scalar Multiplication

Given two vectors **a** and **b**, their sum in both common and tensor notation is,

$$\mathbf{a} + \mathbf{b} = a_i + b_i = (a_1 + b_1)\mathbf{i} + (a_2 + b_2)\mathbf{j} + (a_3 + b_3)\mathbf{k} \tag{2.6}$$

We often need to multiply vectors by scalars. Common and tensor notation for this operation for the example of scaling vector $\mathbf{v} = v_i$ with scalar ϕ. This is written as,

$$\phi\mathbf{v} = \phi v_i = \phi(v_1\mathbf{i} + v_2\mathbf{j} + v_3\mathbf{k}) = \phi v_1\mathbf{i} + \phi v_2\mathbf{j} + \phi v_3\mathbf{k} \tag{2.7}$$

2.3.2 Dot Product

The dot product of two vectors **u** and **v** is defined as,

$$\mathbf{u} \cdot \mathbf{v} = |\mathbf{u}||\mathbf{v}| \cos\theta \tag{2.8}$$

where $|\cdot|$ is the magnitude operator and θ is the angle between vectors **u** and **v**. Starting from first principles,

$$\mathbf{u} \cdot \mathbf{v} = (u_1\mathbf{i} + u_2\mathbf{j} + u_3\mathbf{k}) \cdot (v_1\mathbf{i} + v_2\mathbf{j} + v_3\mathbf{k}) \tag{2.9}$$

$$= u_1v_1\mathbf{i}\cdot\mathbf{i} + u_1v_2\mathbf{i}\cdot\mathbf{j} + u_1v_3\mathbf{i}\cdot\mathbf{k} \tag{2.10}$$

$$+ u_2v_1\mathbf{j}\cdot\mathbf{i} + u_2v_2\mathbf{j}\cdot\mathbf{j} + u_2v_3\mathbf{j}\cdot\mathbf{k}$$

$$+ u_3v_1\mathbf{k}\cdot\mathbf{i} + u_3v_2\mathbf{k}\cdot\mathbf{j} + u_3v_3\mathbf{k}\cdot\mathbf{k}$$

The unit vectors are orthonormal; therefore, the angle between them is $\theta = 0$ and each other is $\theta = \pi/2$.[1] Using the definition in Eq. (2.8), we can write,

$$\mathbf{i}\cdot\mathbf{i} = \mathbf{j}\cdot\mathbf{j} = \mathbf{k}\cdot\mathbf{k} = 1 \tag{2.11}$$

$$\mathbf{i}\cdot\mathbf{j} = \mathbf{i}\cdot\mathbf{k} = \mathbf{j}\cdot\mathbf{i} = \mathbf{j}\cdot\mathbf{k}$$

$$= \mathbf{k}\cdot\mathbf{i} = \mathbf{k}\cdot\mathbf{j} = 0 \tag{2.12}$$

because $\cos(0) = 1$ and $\cos\pi/2 = 0$. Using this development, the dot product can be written more simply as,

$$\mathbf{u} \cdot \mathbf{v} = u_1v_1 + u_2v_2 + u_3v_3 \tag{2.13}$$

The dot product can be written as a summation

$$\mathbf{u} \cdot \mathbf{v} = \sum_{i=1}^{3} u_iv_i = u_1v_1 + u_2v_2 + u_3v_3 \tag{2.14}$$

We can now use tensor notation to simplify our notation further,

[1] For example, the angle between **i** and **i** is $\theta = 0$ and **i** and **j** is $\theta = \pi/2$.

$$\mathbf{u} \cdot \mathbf{v} = \sum_{i=1}^{3} u_i v_i = u_i v_i \tag{2.15}$$

Note how tensor notation can be used to imply summation. The general rule is whenever repeated indices occur in a product there is a summation. In this way, we can avoid excessive use of the summation sign $\sum$. As we will see, this is very useful for more complex vector operations.

To demonstrate this further, we will examine the magnitude squared of a vector. Given the vector $\mathbf{v}$ in both common and tensor notation,

$$\mathbf{v} = v_i = (u, v, w) = u\mathbf{i} + v\mathbf{j} + w\mathbf{k} \tag{2.16}$$

The magnitude squared of $\mathbf{v} = v_i$ is the dot product of itself. In common and tensor notation,

$$\mathbf{v} \cdot \mathbf{v} = |\mathbf{v}|^2 = v_i v_i = v_i^2 = \sum_{i=1}^{3} v_i^2 = u^2 + v^2 + w^2 \tag{2.17}$$

To find the magnitude, take the square root of the magnitude squared,

$$|\mathbf{v}| = \sqrt{v_i^2} \tag{2.18}$$

Another important concept is the *unit vector*, $\mathbf{e_v}$. Consider a vector $\mathbf{v}$. The unit vector of $\mathbf{v}$ is a vector with unity magnitude aligned with $\mathbf{v}$. It is defined as,

$$\mathbf{e_v} = \frac{\mathbf{v}}{|\mathbf{v}|} \tag{2.19}$$

2.3.3 Cross Product

The cross product of two vectors $\mathbf{u}$ and $\mathbf{v}$ is defined as,

$$\mathbf{w} = \mathbf{u} \times \mathbf{v} \tag{2.20}$$

The dot product is a scalar, while the cross product is a vector. Further, the cross product is a vector that is perpendicular to the plane formed by the original vectors that form the product, with the direction of this vector determined by the right hand rule.

The cross product magnitude is,

$$|\mathbf{w}| = |\mathbf{u} \times \mathbf{v}| = |\mathbf{u}||\mathbf{v}| \sin\theta \tag{2.21}$$

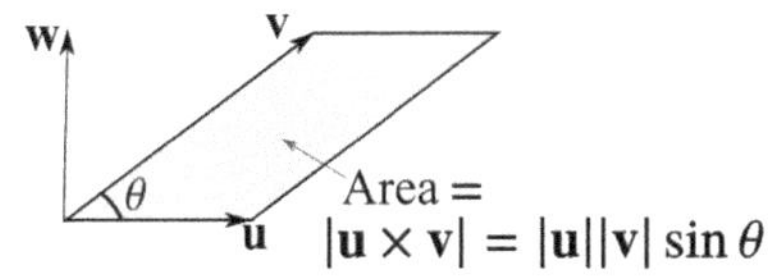

Fig. 2.3 Cross product illustrating the formation of $\mathbf{w} = \mathbf{u} \times \mathbf{v}$. The direction of $\mathbf{w}$ is set by the right hand rule. Note how the magnitude of the cross product is the area of the parallelogram formed by vectors $\mathbf{u}$ and $\mathbf{v}$

where θ is the angle between vectors $\mathbf{u}$ and $\mathbf{v}$. Figure 2.3 describes the geometry of the cross product and shows that the magnitude is actually the area of the parallelogram formed by the vectors $\mathbf{u}$ and $\mathbf{v}$. Figure 2.3 also shows that $\mathbf{w}$ is perpendicular to the plane formed by $\mathbf{u}$ and $\mathbf{v}$. Notice that the direction of $\mathbf{w}$ is set by the right hand rule.

The cross product can be calculated using the determinant method,

$$\mathbf{u} \times \mathbf{v} = \begin{vmatrix} \mathbf{i} & \mathbf{j} & \mathbf{k} \\ u_1 & u_2 & u_3 \\ v_1 & v_2 & v_3 \end{vmatrix} = \begin{bmatrix} u_2v_3 - u_3v_2 \\ u_3v_1 - u_1v_3 \\ u_1v_2 - u_2v_1 \end{bmatrix} = (u_2v_3 - u_3v_2)\mathbf{i} + (u_3v_1 - u_1v_3)\mathbf{j} + (u_1v_2 - u_2v_1)\mathbf{k} \tag{2.22}$$

In tensor notation, the cross product is written,

$$\mathbf{w} = \mathbf{u} \times \mathbf{v} = w_i = \epsilon_{ijk} u_j v_k \tag{2.23}$$

where ϵ_{ijk} is the permutation symbol, defined so that:

- $\epsilon_{ijk} = 1$ if (ijk) is an even (cyclic) permutation of 123: $\epsilon_{123} = \epsilon_{231} = \epsilon_{312} = 1$
- $\epsilon_{ijk} = -1$ if (ijk) is an odd (non-cyclic) permutation of 123: $\epsilon_{213} = \epsilon_{321} = \epsilon_{132} = -1$
- $\epsilon_{ijk} = 0$ otherwise, for example: $\epsilon_{111} = \epsilon_{221} = 0$, etc.

2.3.4 Scalar Triple Product

The scalar triple product is defined as,

$$\mathbf{a} \cdot (\mathbf{b} \times \mathbf{c}) = \begin{vmatrix} a_1 & a_2 & a_3 \\ b_1 & b_2 & b_3 \\ c_1 & c_2 & c_3 \end{vmatrix} \tag{2.24}$$

which is useful for calculating the volume of a parallelepiped formed by $\mathbf{a}$, $\mathbf{b}$ and $\mathbf{c}$.

In tensor notation,

$$\mathbf{a} \cdot (\mathbf{b} \times \mathbf{c}) = \epsilon_{ijk} a_i b_j c_k \tag{2.25}$$

2.4 The 'del' Operator

The 'del' or 'nabla' operator [1, 2, 4] is used to obtain partial derivatives in vector calculus. It is defined in cartesian coordinates as,

$$\nabla = \frac{\partial}{\partial x}\mathbf{i} + \frac{\partial}{\partial y}\mathbf{j} + \frac{\partial}{\partial z}\mathbf{k} \tag{2.26}$$

This is an extremely powerful and useful operator for our work in flow noise. For completeness, the 'del' operator in cylindrical coordinates (r, θ, z) is,

$$\nabla = \frac{\partial}{\partial r}\hat{\mathbf{r}} + \frac{1}{r}\frac{\partial}{\partial \theta}\hat{\boldsymbol{\theta}} + \frac{\partial}{\partial z}\hat{\mathbf{z}} \tag{2.27}$$

where $(\hat{\mathbf{r}}, \hat{\boldsymbol{\theta}}, \hat{\mathbf{z}})$ are the unit vectors in the (r, θ, z) directions, respectively.

In spherical coordinates (r, θ, ϕ), the 'del' operator is,

$$\nabla = \frac{\partial}{\partial r}\hat{\mathbf{r}} + \frac{1}{r}\frac{\partial}{\partial \theta}\hat{\boldsymbol{\theta}} + \frac{1}{r\sin\theta}\frac{\partial}{\partial z}\hat{\boldsymbol{\phi}} \tag{2.28}$$

where $(\hat{\mathbf{r}}, \hat{\boldsymbol{\theta}}, \hat{\boldsymbol{\phi}})$ are the unit vectors in the (r, θ, ϕ) directions, respectively.

2.4.1 Gradient

The gradient of a scalar field (for example $f = f(x, y, z)$ in cartesian coordinates) is a vector field defined as,

$$\nabla f = \frac{\partial f}{\partial x}\mathbf{i} + \frac{\partial f}{\partial y}\mathbf{j} + \frac{\partial f}{\partial z}\mathbf{k} \tag{2.29}$$

thus the gradient of a scalar field is vector whose components are the gradients of the field aligned with each unit vector direction. As an example, consider the two-dimensional field,

$$p(x, y) = e^{-(x/4-y/4)^2} \tag{2.30}$$

Thus, the gradient is,

$$\begin{aligned}\nabla p &= \frac{\partial p}{\partial x}\mathbf{i} + \frac{\partial p}{\partial y}\mathbf{j} \\ &= -\frac{1}{8}(x-y)e^{-(x/4-y/4)^2}\,\mathbf{i} + \frac{1}{8}(x-y)e^{-(x/4-y/4)^2}\,\mathbf{j}\end{aligned} \tag{2.31}$$

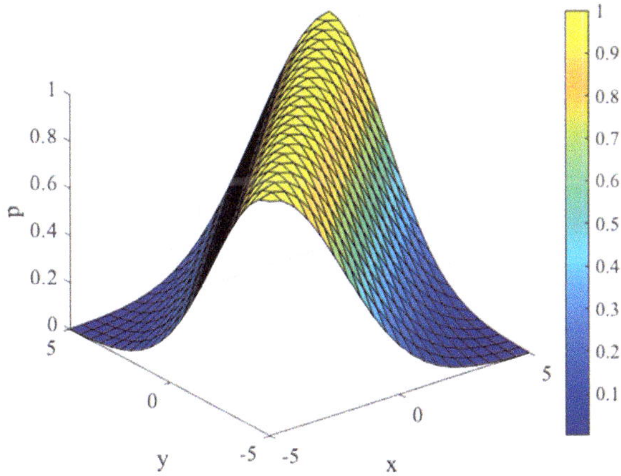

Fig. 2.4 Two-dimensional scalar field $p(x, y) = e^{-(x/4-y/4)^2}$

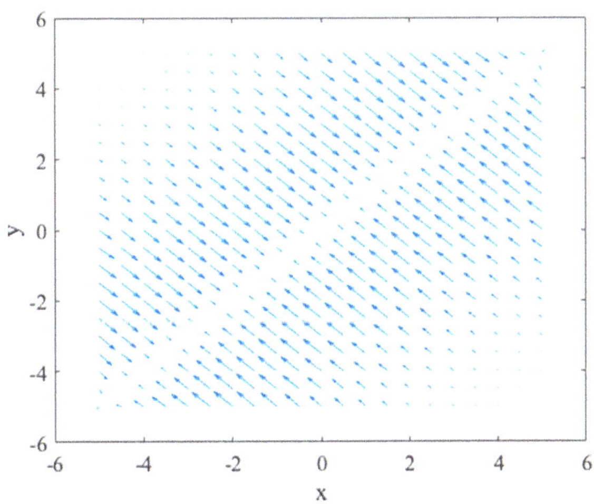

Fig. 2.5 Two dimensional vector field illustrating gradient, $\nabla p = \frac{\partial p}{\partial x}\mathbf{i} + \frac{\partial p}{\partial y}\mathbf{j} = -\frac{1}{8}(x-y)e^{-(x/4-y/4)^2}\,\mathbf{i} + \frac{1}{8}(x-y)e^{-(x/4-y/4)^2}\,\mathbf{j}$. The arrows represent vectors, length is magnitude and the arrow head indicates direction

Figure 2.4 displays Eq. (2.30) while Fig. 2.5 shows the vector field of the gradient of p, shown in Eq. (2.31).

In tensor notation, the gradient of f is written,

$$\nabla f = \frac{\partial f}{\partial x_i} = \frac{\partial f}{\partial x}\mathbf{i} + \frac{\partial f}{\partial y}\mathbf{j} + \frac{\partial f}{\partial z}\mathbf{k} \tag{2.32}$$

2.4.2 *Divergence*

The divergence creates a scalar field from a vector field. Consider the vector field $\mathbf{F} = F_x\mathbf{i} + F_y\mathbf{j} + F_z\mathbf{k}$. The divergence is created by taking the dot product of the del

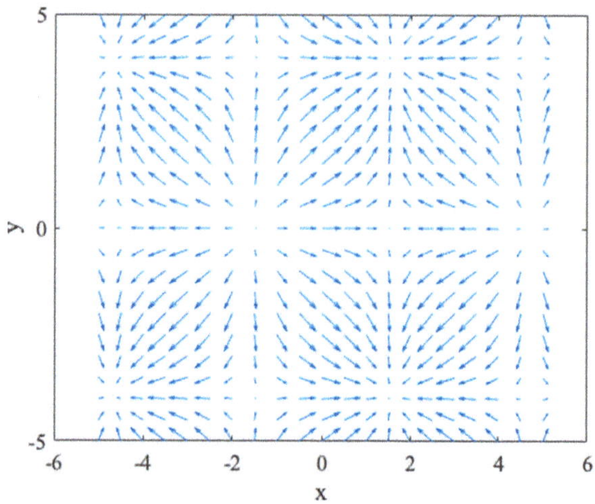

Fig. 2.6 Two-dimensional vector field $\mathbf{F} = \cos(x)\,\mathbf{i} + \sin\left(\frac{4y}{5}\right)\mathbf{j}$

operator and the vector,

$$\nabla \cdot \mathbf{F} = \frac{\partial F_x}{\partial x} + \frac{\partial F_y}{\partial y} + \frac{\partial F_z}{\partial z} \tag{2.33}$$

As an illustration, consider the two-dimensional vector field,

$$\mathbf{F} = \cos(x)\,\mathbf{i} + \sin\left(\frac{4y}{5}\right)\mathbf{j} \tag{2.34}$$

The divergence of this field is a scalar function in two-dimensions,

$$\nabla \cdot \mathbf{F} = -\sin(x) + \frac{4}{5}\cos\left(\frac{4y}{5}\right) \tag{2.35}$$

These fields are shown in Figs. 2.6 and 2.7 respectively.
In Tensor notation, the divergence of vector $\mathbf{F}$ is written,

$$\nabla \cdot \mathbf{F} = \frac{\partial F_i}{\partial x_i} = \frac{\partial F_x}{\partial x} + \frac{\partial F_y}{\partial y} + \frac{\partial F_z}{\partial z} \tag{2.36}$$

2.4.3 Laplacian

The Laplacian is the divergence of the gradient of a scalar field, hence it too is a scalar field. For the scalar field $\phi(x, y, z)$, it is defined as,

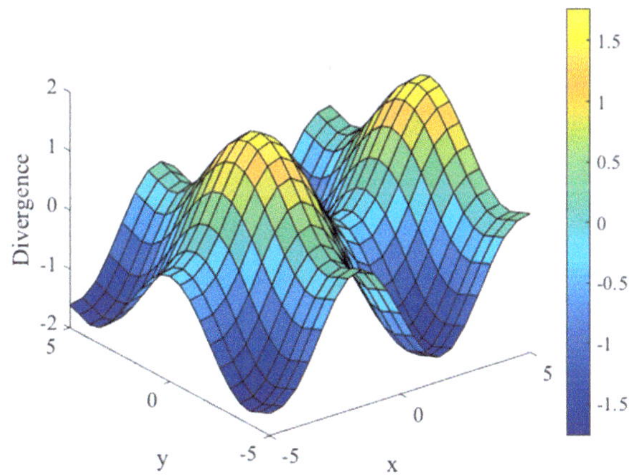

Fig. 2.7 Two dimensional scalar field illustrating divergence, $\nabla \cdot \mathbf{F} = \frac{\partial F_x}{\partial x} + \frac{\partial F_y}{\partial y} = -\sin(x) + \frac{4}{5}\cos\left(\frac{4y}{5}\right)$. The arrows represent vectors, length is magnitude and the arrow head indicates direction

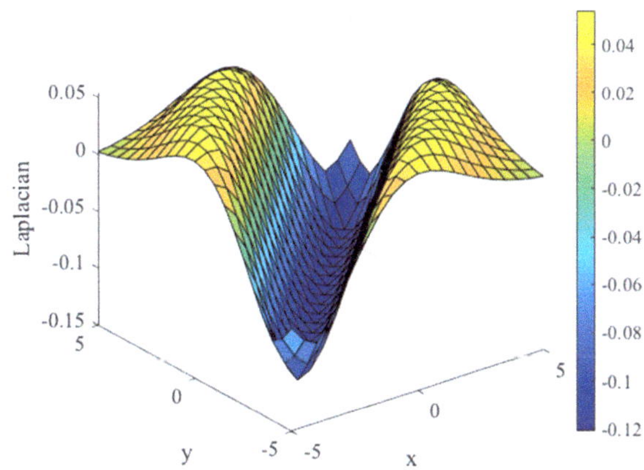

Fig. 2.8 Laplacian of scalar field $p(x, y) = e^{-(x/4-y/4)^2}$, $\nabla^2 p = 2(x/8 - y/8)^2 e^{-(x/4-y/4)^2} - \frac{1}{4}e^{-(x/4-y/4)^2}$

$$\nabla^2 \phi = \nabla \cdot (\nabla \phi) = \frac{\partial^2 \phi}{\partial x^2} + \frac{\partial^2 \phi}{\partial y^2} + \frac{\partial^2 \phi}{\partial z^2} \tag{2.37}$$

For the example used for the gradient (Sect. 2.4.1), the Laplacian of the scalar p (Eq. (2.30)) is shown in Fig. 2.8.

The *vector* Laplacian (or the Laplacian of a vector field) is defined below for the cartesian vector $\mathbf{u}(x, y, z) = u\mathbf{i} + v\mathbf{j} + w\mathbf{k}$,

$$\nabla^2 \mathbf{u} = \frac{\partial^2 u}{\partial x^2}\mathbf{i} + \frac{\partial^2 v}{\partial y^2}\mathbf{j} + \frac{\partial^2 w}{\partial z^2}\mathbf{k} \tag{2.38}$$

In tensor notation, the Laplacian of scalar field ϕ is written,

$$\nabla^2 \phi = \frac{\partial^2 \phi}{\partial x_i^2} = \frac{\partial^2 \phi}{\partial x^2} + \frac{\partial^2 \phi}{\partial y^2} + \frac{\partial^2 \phi}{\partial z^2} \tag{2.39}$$

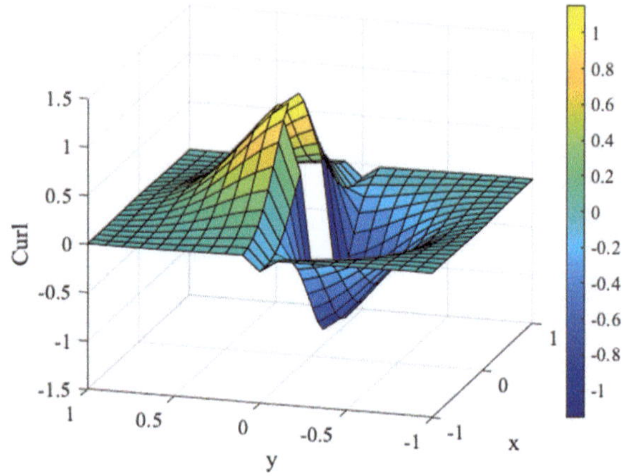

Fig. 2.9 Curl of two-dimensional vector field (Eq. (2.41))

2.4.4 *Curl*

The cross product of the del operator with a vector is known as the curl of that vector. Thus, the curl of the vector $\mathbf{u} = u\mathbf{i} + v\mathbf{j} + w\mathbf{k}$ is defined,

$$\nabla \times \mathbf{u} = \begin{vmatrix} \mathbf{i} & \mathbf{j} & \mathbf{k} \\ \frac{\partial}{\partial x} & \frac{\partial}{\partial y} & \frac{\partial}{\partial z} \\ u & v & w \end{vmatrix} = \begin{bmatrix} \frac{\partial w}{\partial y} - \frac{\partial v}{\partial z} \\ \frac{\partial u}{\partial z} - \frac{\partial w}{\partial x} \\ \frac{\partial v}{\partial x} - \frac{\partial u}{\partial y} \end{bmatrix} = \left(\frac{\partial w}{\partial y} - \frac{\partial v}{\partial z}\right)\mathbf{i} + \left(\frac{\partial u}{\partial z} - \frac{\partial w}{\partial x}\right)\mathbf{j} + \left(\frac{\partial v}{\partial x} - \frac{\partial u}{\partial y}\right)\mathbf{k} \tag{2.40}$$

As an example, consider the two-dimensional cartesian vector,

$$\mathbf{u} = -V_\theta \cos\theta \mathbf{i} + V_\theta \sin\theta \mathbf{j} \tag{2.41}$$

where

$$V_\theta = \frac{1}{2\pi r}\left(1 - e^{-4r^2}\right) \tag{2.42}$$

and $r = \sqrt{x^2 + y^2}$ in a cartesian (x, y) coordinate system.[2] In two dimensions, the curl becomes a single component vector,

$$\nabla \times \mathbf{u} = \left(\frac{\partial v}{\partial x} - \frac{\partial u}{\partial y}\right)\mathbf{k} \tag{2.43}$$

the magnitude of which can be handled like a scalar field. Thus, the magnitude of the two-dimensional curl of Eq. (2.41) is shown in Fig. 2.9.

The curl of cartesian vector $\mathbf{u}$ in tensor notation is written,

---

[2] The reader may recoginse this as a form of the Lamb-Oseen vortex equation.

$$\nabla \times \mathbf{u} = \epsilon_{ijk} \frac{\partial u_k}{\partial x_j} \tag{2.44}$$

2.4.5 *Kronecker Delta Function*

The Kronecker delta function is defined as,

$$\delta_{ij} = \begin{cases} 1, \; i = j \\ 0, \; i \neq j \end{cases} \tag{2.45}$$

It is useful for converting scalars to tensors,

$$p\delta_{ij} = \begin{bmatrix} p & 0 & 0 \\ 0 & p & 0 \\ 0 & 0 & p \end{bmatrix} \tag{2.46}$$

This way, scalars can be included in tensor operations.

2.5 Integrals and Theorems

Here we define line, surface and volume integrals. These are related through *theorems*, which are summarised below.

2.5.1 *Line Integrals*

Consider $F(\mathbf{x})$, a scalar field in three-dimensional space. If C is simple curve, such as that shown in Fig. 2.10, described by $\mathbf{x}(t) = x_i(t)$ where $a \leq t \leq b$, then the line integral of F from a to b is defined as,

$$\int_C F(\mathbf{x}) \, \mathrm{d}t = \int_a^b F(x_i(t)) \, \mathrm{d}t \tag{2.47}$$

recalling that the vector $\mathbf{x} = x_i = (x_1, x_2, x_3)$. If the order of the integration is reversed, then it can be expected that the magnitude of the integral will remain the same, but opposite in sign. If the positions of a and b are the same ($x_i(a) = x_i(b)$), then the curve C is closed (see Fig. 2.11), and the integral is written,

$$\oint_C F(\mathbf{x}) \, \mathrm{d}t \tag{2.48}$$

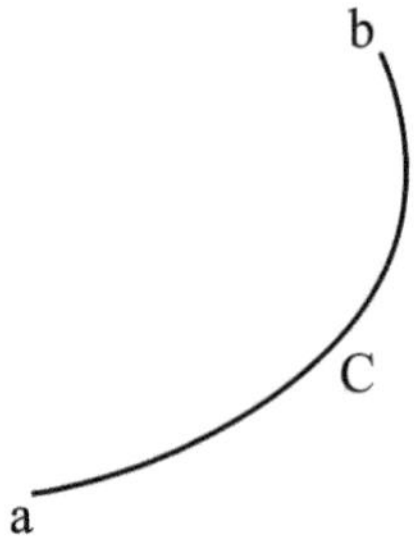

Fig. 2.10 Curve in space passing through either a scalar or vector field

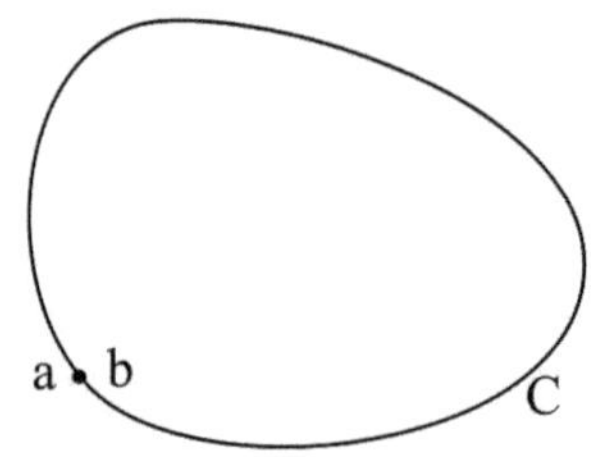

Fig. 2.11 Closed curve in space passing through either a scalar or vector field

where anti-clockwise is considered the positive direction of integration.

Consider now a three-dimensional vector field $\mathbf{A}(\mathbf{x})$. If $\mathbf{t}$ is the unit tangent vector of curve C (as in Fig. 2.11), then $\mathbf{A} \cdot \mathbf{t}$ is the projection of the vector field to the tangent of curve C. The closed line integral of vector field $\mathbf{A}$ over closed curve C is written,

$$\oint_C \mathbf{A} \cdot \mathbf{t}\, \mathrm{d}s \tag{2.49}$$

where s is a parameter describing the position around curve C and $\mathrm{d}s$ is a small increment along that curve. Note that,

$$\frac{\mathrm{d}x_i}{\mathrm{d}s} = t_i \tag{2.50}$$

Thus, Eq. (2.49) could also be written,

$$\oint_C A_i\, \mathrm{d}x_i = \oint_C \mathbf{A} \cdot \mathrm{d}\mathbf{x} \tag{2.51}$$

2.5.2 *Surface Integrals*

We now extend the line integral concept to surfaces. First, consider an open surface S, bounded by closed curve C, as shown in Fig. 2.12. Consider also the incremental area $\mathrm{d}S$ whose outward normal is the unit vector $\mathbf{n}$, when combined they form the

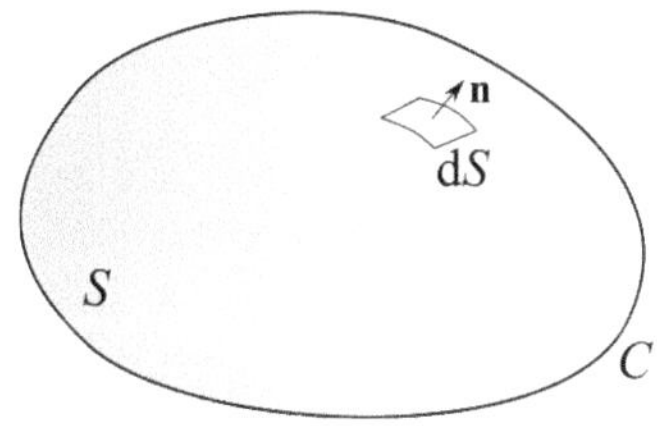

Fig. 2.12 Open surface in space passing through either a scalar or vector field

elemental area vector $\mathbf{dS}$. If we wish to evaluate the integral of a scalar (for example, p) over the open area S, we use the surface integral,

$$\int_S p\,\mathbf{dS} = \int_S p\,\mathrm{d}S_i \tag{2.52}$$

The surface integral of the vector field $\mathbf{F}$ can be evaluated as a dot or cross product. The surface integral using the dot product is,

$$\int_S \mathbf{F} \cdot \mathbf{dS} = \int_S F_i \mathrm{d}S_i \tag{2.53}$$

The result is a scalar.

The surface integral using the cross product is,

$$\int_S \mathbf{F} \times \mathbf{dS} = \int_S \epsilon_{ijk} F_j \mathrm{d}S_k \tag{2.54}$$

The result this time is a vector.

If the surface is closed (such as a sphere), these integrals can be written in the following format,

$$\oint_S p\,\mathbf{dS} = \oint_S p\,\mathrm{d}S_i \tag{2.55}$$

$$\oint_S \mathbf{F} \cdot \mathbf{dS} = \oint_S F_i \mathrm{d}S_i \tag{2.56}$$

$$\oint_S \mathbf{F} \times \mathbf{dS} = \oint_S \epsilon_{ijk} F_j \mathrm{d}S_k \tag{2.57}$$

2.5.3 Volume Integrals

If we consider a volume V, we can define volume integrals in a similar manner to Sect. 2.5.2. For the same scalar and vector fields, the volume integrals are,

$$\oint_V p \, \mathrm{d}V \tag{2.58}$$

$$\oint_V \mathbf{F} \, \mathrm{d}V \tag{2.59}$$

where we note that the elemental volume $\mathrm{d}V$ is a scalar quantity. In a three-dimensional cartesian coordinate system, $\mathrm{d}V = \mathrm{d}x \, \mathrm{d}y \, \mathrm{d}z$.

2.5.4 *Theorems*

Three important theorems are regularly used to relate line, surface and volume integrals with each other.

If we have a volume V with surface S, the surface and volume integrals of vector $\mathbf{F}$ are related by the divergence theorem (or Gauss' theorem),

$$\oint_S \mathbf{F} \cdot \mathbf{dS} = \oint_V (\nabla \cdot \mathbf{F}) \, \mathrm{d}V \tag{2.60}$$

Consider an open surface S bounded by curve C. Stokes theorem relates the line integral of vector field $\mathbf{F}$ over C with the surface integral of $\mathbf{F}$ over S,

$$\oint_C \mathbf{F} \cdot \mathbf{dS} = \int_S (\nabla \times \mathbf{F}) \cdot \mathbf{dS} \tag{2.61}$$

For a scalar field p, the gradient theorem relates the surface and volume integrals as,

$$\oint_S p \, \mathbf{dS} = \oint_V \nabla p \, \mathrm{d}V \tag{2.62}$$

2.6 Dirac Delta Function

The Dirac delta function is a generalised function that is very helpful for the solution of dynamic problems. It is sometimes defined as the unit impulse function. Its definition is,

$$\delta(t - \tau) = \begin{cases} \infty, & \text{if } t = \tau \\ 0, & \text{otherwise} \end{cases} \tag{2.63}$$

and

$$\int_0^\infty \delta(t - \tau) \, \mathrm{d}t = 1 \tag{2.64}$$

While we will make use of the Dirac function later in the book, one use can be described by the convolution of function $G(t)$ with the Dirac delta,

$$\int_{-\infty}^{\infty} G(t)\delta(t-\tau)\,\mathrm{d}t = G(\tau) \tag{2.65}$$

which can also be considered a definition of the Dirac delta [3].

2.7 Summary

This chapter presented a very brief overview of vector calculus that can be used to support reading and learning in the remainder of this book. Scalars, vectors and tensors were defined, along with tensor notation. Basic operations, including the del operator were covered. Line integrals, theorems and delta functions were explained at the end of the chapter.

References

1. Anderson J (2017) Fundamentals of aerodynamics, vol 6th. McGraw Hill, New York
2. Aris R (1989) Vectors, tensors and the basic equations of fluid mechanics. Dover Publications, New York
3. Blake WK (2017) Mechanics of flow-induced sound and vibration, vol 1: General concepts and elementary sources, 2nd edn. Academic press, London
4. Kreyszig E (1988) Advanced engineering mathematics, 6th edn. Wiley
5. Moukalled F, Mangani L, Darwish M (2016) The finite volume method in computational fluid dynamics, an advanced introduction with OpenFOAM and Matlab. Springer, Heidelberg

Chapter 3
Spectral Analysis

Abstract This chapter provides fundamental information concerning basic mathematical tools used to interpret signals that vary in space and time. Approximations to trigonometric functions are used to introduce the chapter. This is followed by an introduction to Bessel's equation and Bessel functions. Exponential and complex harmonic functions are defined, including the form that is used in the rest of this textbook. Fourier series are introduced and the reader is shown how they can be used to represent any periodic signal. The Fourier transform is also defined and used to express continuous, random signals in the frequency domain. Finally, the one-dimensional wavenumber spectrum is defined for spatially varying signals.

3.1 Introduction

This chapter introduces key concepts in spectral analysis. These are fundamental to the understanding and analysis of flow-noise problems. Classical acoustics consists of the decomposition of a time-varying pressure signal into a spectrum representing energy in each frequency band. The study of turbulence also requires the decomposition of velocity signals into frequency or wave number spectra for their analysis. Further, we often want to understand the link, or *correlation* between signals (acoustic, turbulent pressure, etc.) in order to deduce the origin of flow noise (the source). This chapter provides a grounding in these areas.

The chapter begins with approximations to trigonmetric functions. This is followed by an introduction to Bessel's equation and Bessel functions, often used to describe mode shapes in ducts. Exponential and complex harmonic functions are covered in Sect. 3.4. Importantly, the convention used for the exponential function is defined here.

Fourier Series are presented in Sect. 3.5 which leads to the Fourier Transform in Sect. 3.6 (in both time and space). These last two sections are fundamental to understanding many aspects of acoustics and turbulence and although quite short, are important.

C. Doolan and D. Moreau, *Flow Noise*,
https://doi.org/10.1007/978-981-19-2484-2_3

3.2 Trigonometric Functions and Their Approximation

It is, of course, assumed that the reader is familiar with sine and cosine trigonometric functions. It is not often discussed how these functions can be approximated or evaluated numerically; however, they are included here for completeness and for those wanting to simplify approximate models of flow noise systems. A straightforward proof [5] shows that the trigonometric function can be evaluated using the following series,

$$\cos x = 1 - \frac{x^2}{2} + \frac{x^4}{24} - \frac{x^6}{720} + \cdots + \frac{x^n}{n!} + \cdots \tag{3.1}$$

$$\sin x = 1 - \frac{x^3}{6} + \frac{x^5}{120} - \frac{x^7}{5040} + \cdots + \frac{x^m}{m!} + \cdots \tag{3.2}$$

where $n = 2, 4, 6, 8, \ldots$ are even integers and $m = 1, 3, 5, 7, \ldots$ are odd integers.

Figure 3.1 and 3.2 compare Eqs. (3.1) and (3.2) for various levels of summation (indicated by n and m) with the exact solutions for cosine and sine, respectively. Reasonable approximations are possible for small values of x/π using small values of n and m.

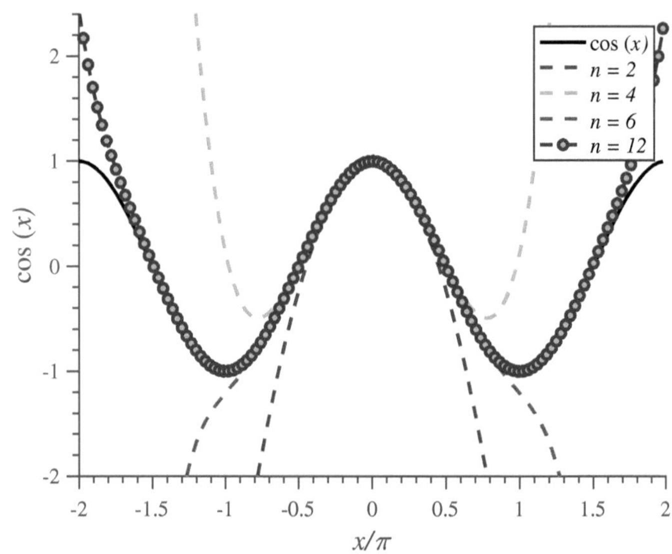

Fig. 3.1 Illustration of approximation of the cosine function using Eq. (3.1) and various values of n

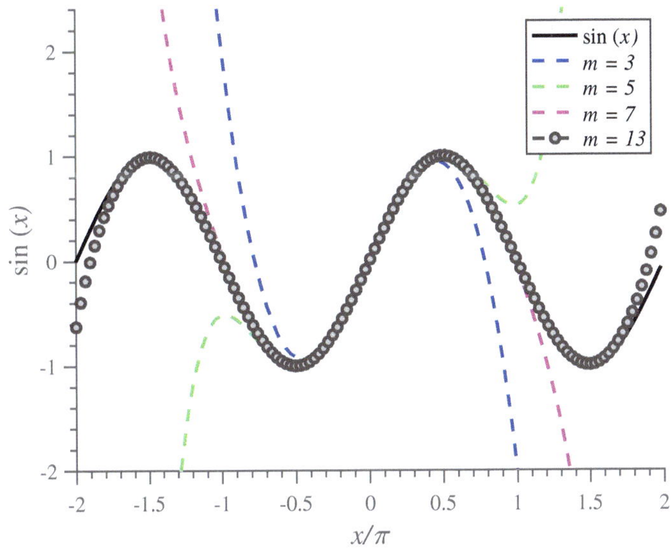

Fig. 3.2 Illustration of approximation of the sine function using Eq. (3.2) and various values of m

3.3 Bessel's Equation and Bessel Functions

Bessel's equation is a useful tool for describing acoustic motion, usually for waves in cylindrical geometries (for example, ducts). Bessel's equation is,

$$\frac{\partial^2 y}{\partial x^2} + \frac{1}{x}\frac{\partial y}{\partial x} + \left(1 - \frac{n^2}{x^2}\right) y = 0 \tag{3.3}$$

where n is a given number called the order of the Bessel equation. If n is not an integer and $x \neq 0$, then the solution to Eq. (3.3) [4] is,

$$y(x) = c_1 J_n(x) + c_2 J_{-n}(x) \tag{3.4}$$

where J_n is the Bessel function of the first kind of order n. The coefficients c_1 and c_2 are arbitrary and are set to suit the particular problem the Bessel function is used to solve. The Bessel function of the first kind of order n is defined [4],

$$J_n = x^n \sum_{m=0}^{\infty} \frac{(-1)^m x^{2m}}{2^{2m+n} m!(n+m)!} \tag{3.5}$$

and we state without proof,

$$J_{-n}(x) = (-1)^n J_n(x) \tag{3.6}$$

If n is an integer, the solution of Bessel's equation is,

$$y(x) = c_1 J_n(x) + c_2 Y_n(x) \tag{3.7}$$

where $Y_n(x)$ is the Bessel function of the second kind. For the case where $n = 0$, we have the Bessel function of the second kind of zero order,

$$Y_0(x) = \frac{2}{\pi}\left[J_0(x)\left(\ln\frac{x}{2} + \gamma\right) + \sum_{m=1}^{\infty}\frac{(-1)^{m-1}h_m}{2^{2m}(m!)^2}x^{2m}\right] \tag{3.8}$$

where $\gamma = 0.577215$ is the Euler constant and h_m is defined as,

$$h_m = 1 + \frac{1}{2} + \cdots + \frac{1}{m} \tag{3.9}$$

For $n = 1, 2, 3, \ldots$, we have the Bessel function of the second kind of order n,

$$Y_n(x) = \frac{1}{\sin n\pi}\left[J_n(x)\cos n\pi - J_{-n}(x)\right] \tag{3.10}$$

It should be noted that Eq. (3.10) is valid for all real values of n (including non-integers). For integer values of $n = p$, the Bessel function of the second kind of order n is found by taking the limit $n \to p$, and the reader is referred to Ref. [4] for complete details.

Examples of Bessel functions of the first and second kind are shown in Figs. 3.3 and 3.4 respectively.

We conclude this section by briefly considering the situation where complex solutions of Bessel's equation are needed. These are known as the Hankel functions, which are [4],

$$H_n^{(1)} = J_n(x) + iY_n(x) \tag{3.11}$$

$$H_n^{(2)} = J_n(x) - iY_n(x) \tag{3.12}$$

where $H_n^{(1)}$ is the Hankel function of the first kind of order n and $H_n^{(2)}$ is the Hankel function of the first kind of order n. They are sometimes called Bessel functions of the third kind.

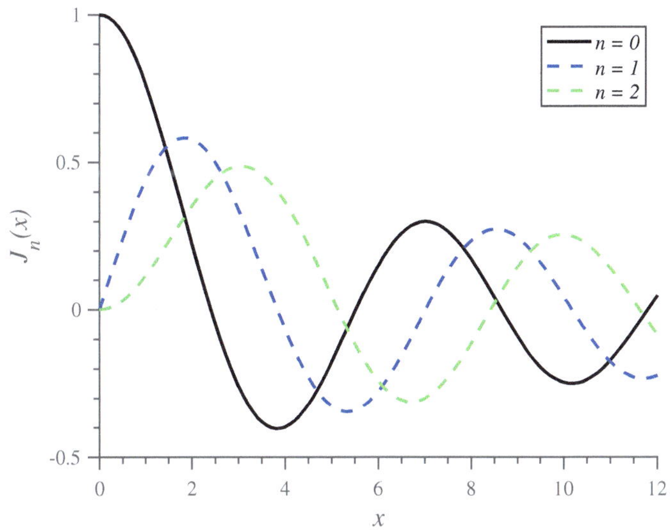

Fig. 3.3 Bessel functions of the first kind (Eq. 3.5), for various values of n

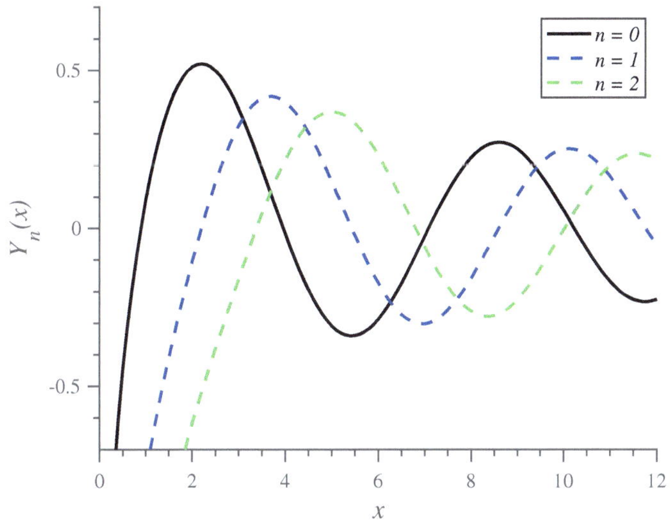

Fig. 3.4 Bessel functions of the second kind (Eq. 3.10), for various values of n

3.4 Exponential and Complex Harmonic Functions

The exponential form of the trigonometric functions is heavily used in flow-induced noise and aeroacoustics. This book uses the following definitions,

$$e^{-ix} = \cos x - i \sin x \tag{3.13}$$

$$\cos x = \frac{1}{2}(e^{ix} + e^{-ix}) \tag{3.14}$$

$$\sin x = -\frac{1}{2}(e^{ix} - e^{-ix}) \tag{3.15}$$

Hence, any complex number can be written as,

$$\hat{A} = a + ib = |A|e^{i\Phi} \tag{3.16}$$

where the hat denotation $\hat{\cdot}$ means the symbol beneath it is complex.

The magnitude of the complex number is defined,

$$|A| = \sqrt{a^2 + b^2} \tag{3.17}$$

and the phase Φ is defined,

$$\tan \Phi = \frac{b}{a} \tag{3.18}$$

The complex conjugate is defined,

$$\hat{A}^* = |A|e^{-i\Phi} = a - ib \tag{3.19}$$

where the superscript $\cdot^*$ indicates the conjugate. The conjugate is often used to define the magnitude,

$$|A|^2 = \hat{A}\hat{A}^* = a^2 + b^2 \tag{3.20}$$

Oscillating quantities that depend on time (at frequency f and angular frequency $\omega = 2\pi f$) can be expressed using a complex exponential function. For example,

$$\hat{P}(\omega) = \hat{A}e^{-i\omega t} = |A|e^{-i(\omega t - \Phi)} \tag{3.21}$$

where t is time. Thus we use complex notation to incorporate magnitude and phase in our description of time dependent quantities.

3.5 Fourier Series

Periodic signals exactly repeat themselves over a fixed period of time, T_p, known as the period. A periodic signal, $x(t)$, with fundamental frequency $f_0 = \frac{1}{T_p} = \frac{\omega_0}{2\pi}$, can be expressed as a Fourier series,

$$x(t) = \frac{a_0}{2} + \sum_{n=1}^{\infty} [a_n \cos(n\omega_0 t) + b_n \sin(n\omega_0 t)] \tag{3.22}$$

where

$$a_n = \frac{2}{T_p}\int_0^{T_p} x(t) \cos n\omega_0 t \, \mathrm{d}t \qquad n = 0, 1, 2, \ldots \tag{3.23}$$

$$b_n = \frac{2}{T_p}\int_0^{T_p} x(t) \sin n\omega_0 t \, \mathrm{d}t \qquad n = 1, 2, 3, \ldots \tag{3.24}$$

Another convenient version is,

$$x(t) = A_0 + \sum_{n=1}^{\infty} A_n \cos(n\omega_0 t - \Phi_n) \tag{3.25}$$

where

$$A_0 = \frac{a_0}{2} \tag{3.26}$$

$$A_n = \sqrt{a_n^2 + b_n^2} \qquad n = 1, 2, 3, \ldots \tag{3.27}$$

$$\tan(\Phi_n) = \frac{b_n}{a_n} \qquad n = 1, 2, 3, \ldots \tag{3.28}$$

Note that we have defined above the fundamental frequency as ω_0; we now define the *harmonics* as the higher order terms, for $n > 1$, and they occur at frequencies $n\omega_0$ for $n > 1$. It is also important to realise that A_0 is the mean of the signal over one cycle, which can be understood by examining Eqs. (3.23) and (3.26),

$$A_0 = \frac{a_0}{2} = \frac{2}{2T_p}\int_0^{T_p} x(t) \, \mathrm{d}t = \overline{x(t)} \tag{3.29}$$

where $\overline{x(t)}$ represents the mean of $x(t)$.

The Fourier series can be defined in a compact form,

$$x(t) = \sum_{n=-\infty}^{\infty} C_n e^{-in\omega_0 t} \tag{3.30}$$

where

$$C_0 = \frac{a_0}{2} = A_0 \tag{3.31}$$

$$C_n = \frac{1}{2}(a_n + ib_n) \tag{3.32}$$

$$C_{-n} = \frac{1}{2}(a_n - ib_n) = C_n^* \tag{3.33}$$

We now determine the complex amplitudes using,

$$C_n = \frac{1}{T_p}\int_0^{T_p} x(t)e^{in\omega_0 t}\,\mathrm{d}t \tag{3.34}$$

The various definitions of the Fourier series above all say that any complex periodic signal can be represented by the sum of its mean and an infinite number of harmonic components with amplitude A_n and phase Φ_n. An amplitude spectrum is formed by presenting each discrete amplitude as a function of frequency.

As an example, consider the following periodic signal,

$$f(t) = 0.5 + 10\cos(\omega_0 t) + 2\cos(5\omega_0 t) + 7\cos(8\omega_0 t) \tag{3.35}$$

A comparison of the original function (Eq. 3.35) with an approximation using a Fourier series using $n = 10$ harmonics is shown in Fig. 3.5. The original function is closely approximated by the Fourier series.

The amplitude spectrum, the representation of A_n as a function of frequency, is shown in Fig. 3.6. Comparison of the amplitudes in Fig. 3.6 with Eq. (3.25) illustrates how the amplitude spectrum is formed, with frequency components only at $n = 0, 1, 5$ and 8 which correspond exactly with the mean and amplitude of each cosine function. Recall that A_0 occurs at zero frequency and is the mean of the signal $f(t)$.

A more interesting example is the square wave signal. A comparison of a unit amplitude square wave repeating pulse with a Fourier approximation ($n = 10$) is shown in Fig. 3.7. While $n = 10$ harmonic components does not accurately represent the original square wave signal, for our purposes it is useful because it shows how combinations of harmonics are used to approximate general functions. The amplitude spectrum in Fig. 3.8 shows the distribution of the amplitudes at each frequency.

Finally, consider a Gaussian signal,

$$f(t) = e^{-\frac{(t-0.5)^2}{0.0015}} \tag{3.36}$$

which is assumed repeating over $T_p = 1$. The Fourier approximation (using $n = 10$) is shown in Fig. 3.9 and the amplitude spectrum in Fig. 3.10.

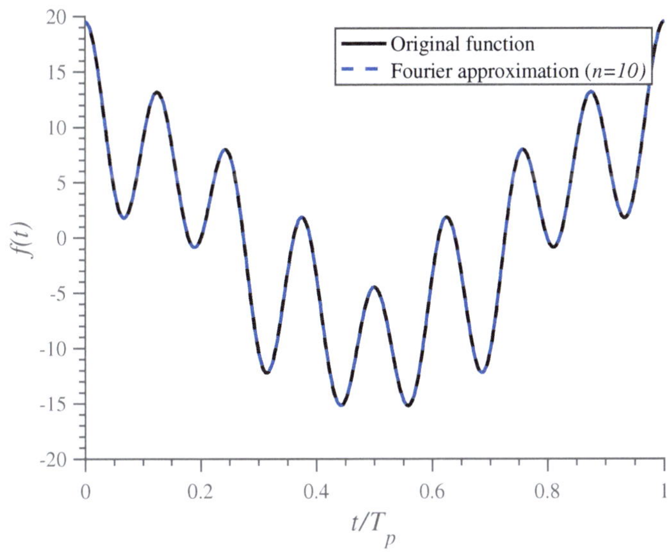

Fig. 3.5 Comparison of a periodic signal (Eq. 3.35) with a Fourier series approximation using $n = 10$ harmonics

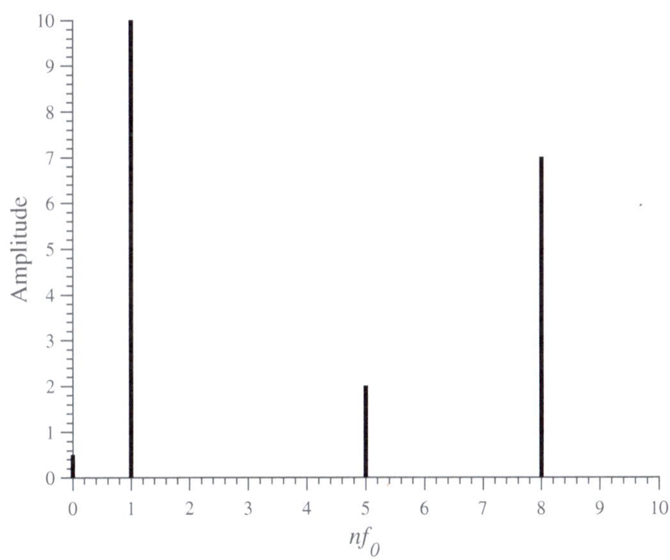

Fig. 3.6 Amplitude spectrum of a periodic signal (Eq. 3.35) using $n = 10$ harmonics

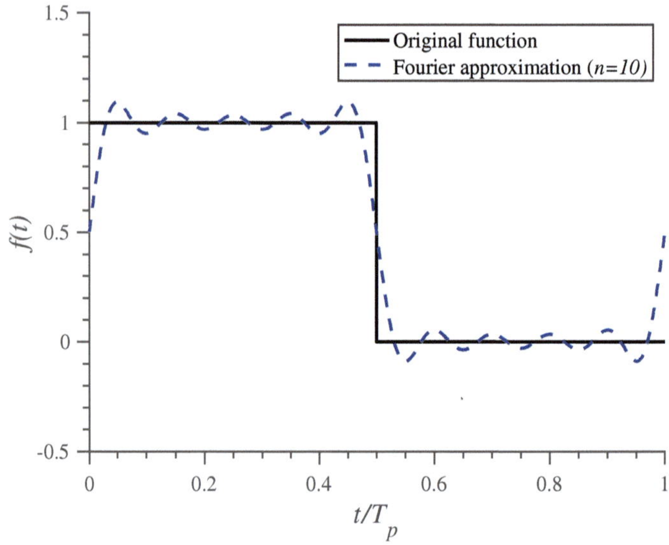

Fig. 3.7 Comparison of a square wave signal with a Fourier series approximation using $n = 10$ harmonics

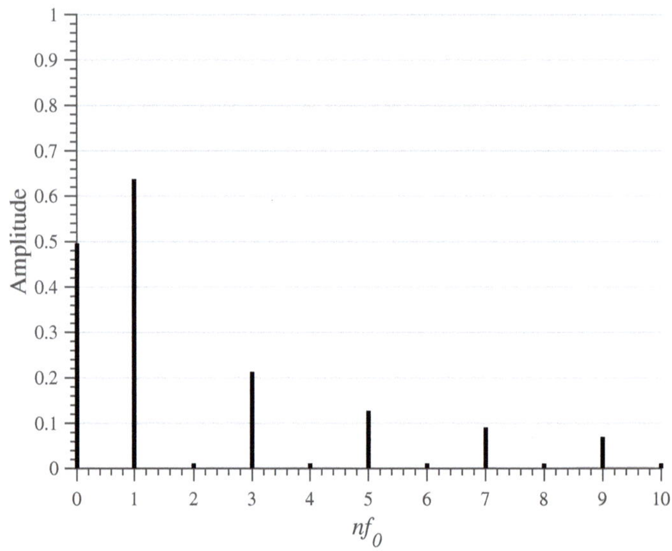

Fig. 3.8 Amplitude spectrum of a square wave signal using $n = 10$ harmonics

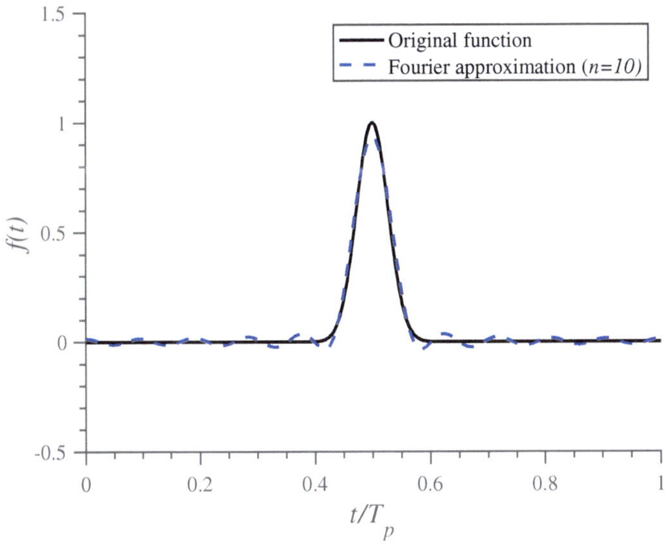

Fig. 3.9 Comparison of a Gaussian signal (Eq. 3.36) with a Fourier series approximation using $n = 10$ harmonics

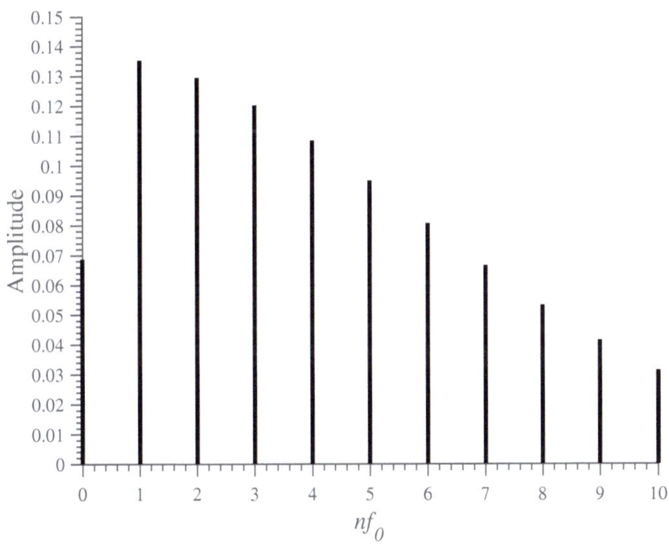

Fig. 3.10 Amplitude spectrum of a Gaussian signal (Eq. 3.36) using $n = 10$ harmonics

3.6 Fourier Transform

For the case where the signal is random and continuous, Fourier analysis can be extended to an integral formulation by assuming that the period T goes to infinity [3]. By doing so, we form the Fourier transform pair. For a given random signal $x(t)$ the definition used in this book is,

$$x(\omega) = \frac{1}{2\pi}\int_{-\infty}^{\infty} x(t)e^{i\omega t}\,\mathrm{d}t \tag{3.37}$$

$$x(t) = \int_{-\infty}^{\infty} x(\omega)e^{-i\omega t}\,\mathrm{d}\omega \tag{3.38}$$

Note that this assumes that the frequency is expressed as the angular frequency $\omega = 2\pi f$ and that the harmonic dependence is expressed using the exponential form $e^{-i\omega t}$. The Fourier transform can be written in other ways [3, for example] if different harmonic dependence is assumed. The convention used in this book is the same as other aeroacoustics [2] and flow noise [1] texts.

The Fourier transform can be applied over space, known as the wavenumber spectrum. In one-dimension, the wavenumber spectrum of the function $f(x_1)$ over dimension x_1 is defined as,

$$f(k_1) = \frac{1}{2\pi}\int_{-\infty}^{\infty} f(x_1)e^{-ik_1x_1}\,\mathrm{d}x_1 \tag{3.39}$$

where k_1 is the wavenumber,

$$k_1 = \frac{2\pi}{\lambda_1} = \frac{\omega}{v_p} \tag{3.40}$$

where λ_1 is the wavelength in the x_1 direction and v_p is the phase velocity. This can be either the speed of sound (c_0, for an acoustic wave) or the convective velocity (U_c, for turbulence).

The inverse of the wavenumber spectrum is,

$$f(x_1) = \int_{-\infty}^{\infty} f(k_1)e^{ik_1x_1}\,\mathrm{d}k_1 \tag{3.41}$$

3.7 Summary

This chapter has provided the tools to understand and analyse complex, temporally and spatially varying signals. The chapter covers trigonometric approximations, Bessel functions, complex harmonic representation of signals, Fourier series and Fourier transforms.

References

1. Blake WK (2017) Mechanics of flow-induced sound and vibration, vol 1: General concepts and elementary sources, 2nd edn. Academic press, London
2. Glegg S, Devenport W (2017) Aeroacoustics of low mach number flows: fundamentals, analysis, and measurement. Academic Press
3. Hansen CH, Doolan CJ, Hansen KL (2017) Wind farm noise: measurement, assessment, and control. Wiley
4. Kreyszig E (1988) Advanced engineering mathematics, 6th edn. Wiley
5. Morse PM, Ingard KU (1986) Theoretical acoustics. Princeton Universty Press, Princeton, NJ, USA

Chapter 4
Fundamental Equations of Fluid Mechanics

Abstract This chapter presents the fundamental equations of fluid mechanics, the building blocks of flow noise analysis. The chapter begins by defining the flow field in terms of spatially and temporally varying scalar and vector field variables. Control volumes are defined next before derivations of the continuity, momentum and energy equations are made.

4.1 Introduction

In this chapter, we introduce and derive the fundamental equations of fluid mechanics. These are the foundations of flow noise and will be used to develop the remaining principles in this book.

The equations of motion are based on the following conservation laws, which are used to derive the continuity, momentum and energy equations:

1. Conservation of mass
2. Conservation of momentum, or Newton's second law of motion; and
3. Conservation of energy.

4.2 Field Specifications

The foundation of fluid mechanics is the *flow field*. Thus, we specify that fluid be described by a number of three-dimensional, time-varying scalar and vector fields. Unless otherwise stated, we assume a cartesian coordinate system, defined in Fig. 4.1.

Figure 4.1 shows the path of a fluid particle through cartesian space as a thick line. We define the main scalar and vector fields below.

C. Doolan and D. Moreau, *Flow Noise*,
https://doi.org/10.1007/978-981-19-2484-2_4

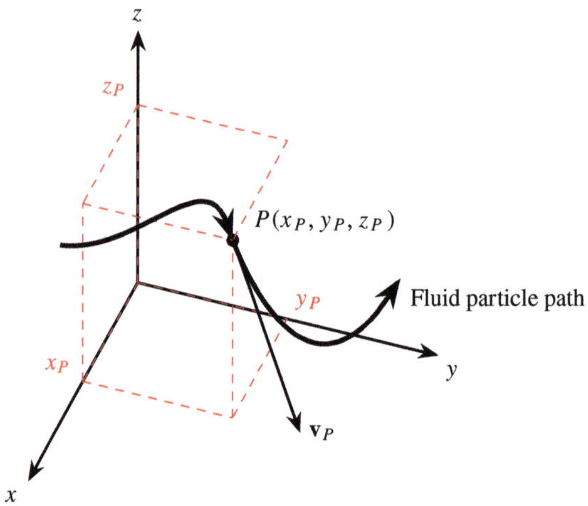

Fig. 4.1 Cartesian coordinate system, defined in (x, y, z)

$$p = p(x, y, z, t) \tag{4.1a}$$
$$\rho = \rho(x, y, z, t) \tag{4.1b}$$
$$T = T(x, y, z, t) \tag{4.1c}$$
$$\mathbf{v} = u\mathbf{i} + v\mathbf{j} + w\mathbf{k} \tag{4.1d}$$

The scalar flow variables we are mainly concerned with are pressure, p; density, ρ and temperature, T. The vector field associated with the velocity of the fluid at each point in space is $\mathbf{v}$. The components of $\mathbf{v}$ are also scalar fields and these are defined below.

$$u = u(x, y, z, t) \tag{4.2a}$$
$$v = v(x, y, z, t) \tag{4.2b}$$
$$w = w(x, y, z, t) \tag{4.2c}$$

Here, u, v and w are the components of $\mathbf{v}$ aligned with the x, y and z axes respectively. Similarly, the symbols **i, j** and **k** represent the unit vectors aligned with the x, y and z axes.

Note that the fields defined above are time-varying (unsteady) as well as spatially varying. They are a complete way of describing the flow field in space and time. Of course, there are almost infinite other field quantities that we can use, and these are mainly combinations or manipulations of the above field variables. Also remember that the thermodynamic state of the fluid can be specified by any two state variables. Three state variables are shown in Eq. 4.1, pressure, p; density, ρ and temperature, T. These state variables are linked via an equation of state, which we will introduce later in this chapter.

Figure 4.1 shows the path of a fluid particle through space. At time t_P, the particle passes through point $P(x_P, y_P, z_P)$, as shown. At this point in time, the velocity vector is shown as $\mathbf{v}_P$, which is a vector tangent to the fluid particle path. For the general case where the flow is completely unsteady and non-uniform (such as in a turbulent flow), it can be expected that at the next instant in time another fluid particle will pass through point P and, because this particle will most likely have its own unique particle path, the velocity vector at point P will be different. Hence, we can expect that at each point in an unsteady, non-uniform flow, the field variables will change with space and time.

4.3 Control Volumes

Our derivation and analysis of the fundamental equations of fluid mechanics will rely on using control volumes. Figure 4.2 illustrates a typical control volume that surrounds a spatially and temporally varying flow field. The flow field has a velocity field represented by vector $\mathbf{v}$, as defined in Eq. (4.1d). The control volume is finite, has volume V, and is enclosed by surface S. We are generally interested in what passes through the boundary of a control volume (referred to as a *flux*, defined later); that is, we want to know what mass, momentum and energy is passing through each point on the surface S at each instant in time. In order to do this analysis, we define the elemental surface area $\mathrm{d}S$, which has an outward pointing normal vector, $\mathbf{n}$.

We use the control volume with the conservation laws (mass, momentum and energy are conserved) to derive the fundamental equations, known as the continuity equation, the momentum equation (with special variant, the Navier Stokes equation), and the energy equation.

4.4 Continuity Equation

The continuity equation is how the conservation of mass is expressed in a form suitable for fluid mechanics. To begin, let's consider the *mass flux* across the elemental area $\mathrm{d}S$, as shown in Fig. 4.2. The mass flux is the flow of mass per unit time across $\mathrm{d}S$. We can express the mass flux in terms of our field variables,

$$\text{Mass flux} = \rho \mathbf{v} \cdot \mathbf{n}\, \mathrm{d}S \tag{4.3}$$

Recall that the dot product $\mathbf{v} \cdot \mathbf{n}$ provides the component of the flow velocity vector $\mathbf{v}$ aligned with the outward normal vector $\mathbf{n}$. The mass flux allows us to keep track of the change in mass within the control volume V with time.

Consider now the mass of fluid contained with V,

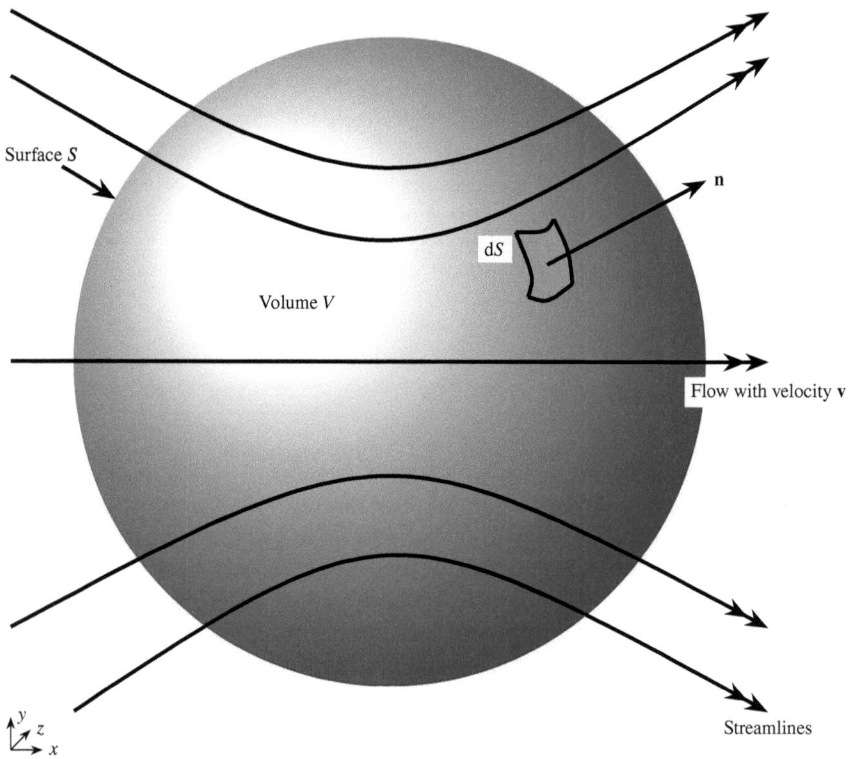

Fig. 4.2 Control volume V, enclosed by surface S placed in a flow field with velocity $\mathbf{v}$ (represented by streamlines) that is flowing through it. Elemental area $\mathrm{d}S$ is shown with outward normal unit vector $\mathbf{n}$

$$\text{Mass of fluid in control volume } V = \int_V \rho \, \mathrm{d}V \tag{4.4}$$

Where the integral $\int_V$ in Eq. (4.4) is a volume integral over volume V.

The fundamental principle of the conservation of mass states that mass cannot be created or destroyed. For the finite control volume, this means that the mass flux across the surface S must equal the time rate of change of mass within volume V. Thus, we can write,

$$\int_S \rho \mathbf{v} \cdot \mathbf{n} \, \mathrm{d}S = -\frac{\partial}{\partial t} \int_V \rho \, \mathrm{d}V = -\int_V \frac{\partial \rho}{\partial t} \mathrm{d}V \tag{4.5}$$

Here, the integral $\int_S$ is the surface integral over surface S. The left hand side of Eq. (4.5) is defined positive for net mass flux outwards from surface S. Therefore, the right hand side of Eq. (4.5) has a negative sign to indicate mass loss due to this net outwards flow. Note also that the derivative $\frac{\partial}{\partial t}$ can be taken into the integral and applied to the first term ρ only because the volume is constant.

Using the divergence theorem, the surface integral on the left hand side of Eq. (4.5) can be converted into a volume integral,

$$\int_S \rho \mathbf{v} \cdot \mathbf{n} \, \mathrm{d}S = \int_V \nabla \cdot (\rho \mathbf{v}) \, \mathrm{d}V \tag{4.6}$$

Hence, Eq. (4.5) becomes,

$$\int_V \frac{\partial \rho}{\partial t} \mathrm{d}V + \int_V \nabla \cdot (\rho \mathbf{v}) \, \mathrm{d}V = 0 \tag{4.7a}$$

$$\int_V \left(\frac{\partial \rho}{\partial t} + \nabla \cdot (\rho \mathbf{v}) \right) \mathrm{d}V = 0 \tag{4.7b}$$

As the control volume V is arbitrary and finite, it means that this result is independent of V. Therefore, the differential form of the continuity equation is,

$$\frac{\partial \rho}{\partial t} + \nabla \cdot (\rho \mathbf{v}) = 0 \tag{4.8}$$

4.5 Momentum Equation

The momentum equation is the expression used in fluid mechanics to describe the principle of conservation of momentum, which is a direct application of Newton's second law: the rate of change of momentum of a body (of fluid in this case) is equal to the sum of forces acting on that body (of fluid). In the case of a fluid, we seek the rate of change of momentum within the control volume, which must be equal to the sum of forces acting on that control volume. As for the continuity equation, we will use the control volume shown in Fig. 4.2 to support our derivation.

The momentum flux out of the control surface can be written,

$$\text{Momentum flux} = (\rho \mathbf{v} \cdot \mathbf{n} \, \mathrm{d}S) \, \mathbf{v} \tag{4.9}$$

As per the previous analysis for the continuity equation, the momentum flux is defined as outwards because we have defined the normal vector $\mathbf{n}$ as outward facing.

The momentum within the control volume V is,

$$\text{Momentum inside control volume } V = \int_V \rho \mathbf{v} \, \mathrm{d}V \tag{4.10}$$

Newton's second law states that the change in momentum within the control volume must be equal to the force $\mathbf{F}$ applied to that volume. The rate of change of momentum must include both the momentum flux across the control volume boundaries plus the rate of change of that volume itself. Thus we can write,

$$\begin{aligned}&\text{Rate of change of momentum inside control volume } V \\ &= -\text{Momentum flux across surface } S + \text{Force}\end{aligned} \tag{4.11a}$$

$$\frac{\partial}{\partial t}\int_V \rho\mathbf{v}\,\mathrm{d}V = -\int_S (\rho\mathbf{v}\cdot\mathbf{n})\,\mathbf{v}\,\mathrm{d}S + \mathbf{F} \tag{4.11b}$$

$$\int_V \frac{\partial\rho\mathbf{v}}{\partial t}\,\mathrm{d}V = -\int_S (\rho\mathbf{v}\cdot\mathbf{n})\,\mathbf{v}\,\mathrm{d}S + \mathbf{F} \tag{4.11c}$$

Note that the elemental area $\mathrm{d}S$ is a scalar and thus can be brought outside the brackets shown in Eq. (4.9).

Before reducing these equations further, the force vector $\mathbf{F}$ acting on the fluid control volume will be defined. There are three types of force that can act on the fluid. The first are the *body forces* that *act at a distance* on the fluid inside V. These include gravitational forces and magnetohydrodynamics forces. The second and third forces *act at the surface*; these are the pressure forces and the viscous forces, respectively.

We define the body force per unit mass vector as $\mathbf{f}$ (so if gravity was the body force, f is the local acceleration due to gravity). The body force acting on the control volume V is,

$$\text{Body force on } V = \int_V \rho\mathbf{f}\,\mathrm{d}V \tag{4.12}$$

Similarly, the pressure force acting on the elemental surface area $\mathrm{d}S$ is,

$$\text{Pressure force on } \mathrm{d}S = -p\mathbf{n}\,\mathrm{d}S \tag{4.13}$$

The negative sign is used because the pressure is assumed to act inwards. Hence the total pressure force acting on surface S is,

$$\text{Pressure force on } S = -\int_S p\mathbf{n}\,\mathrm{d}S \tag{4.14}$$

Finally, we have the viscous forces that act on the surface of the control volume. We will denote these as the vector $\mathbf{F}_{\mathbf{visc}}$. The viscous forces are created by the action of shear within the fluid. They are controlled by the viscosity of the fluid and the velocity gradients that exist within the flow field.

The total force $\mathbf{F}$ is the combination of the body forces, the pressure forces and the viscous forces,

$$\mathbf{F} = \int_V \rho\mathbf{f}\,\mathrm{d}V - \int_S p\mathbf{n}\,\mathrm{d}S + \mathbf{F}_{\mathbf{visc}} \tag{4.15}$$

We now return to the momentum equation as a whole. Substituting Eq. (4.15) into Eq. (4.11c) yields,

$$\int_V \frac{\partial \rho \mathbf{v}}{\partial t}\,\mathrm{d}V = -\int_S (\rho \mathbf{v}\cdot\mathbf{n})\,\mathbf{v}\,\mathrm{d}S + \int_V \rho \mathbf{f}\,\mathrm{d}V - \int_S p\mathbf{n}\,\mathrm{d}S + \mathbf{F}_{\mathbf{visc}} \tag{4.16}$$

Applying the gradient theorem to the pressure force term (Eq. 4.14) allows it to be expressed as a volume integral,

$$\int_S p\mathbf{n}\,\mathrm{d}S = \int_V \nabla p\,\mathrm{d}V \tag{4.17}$$

Similarly, applying the divergence theorem to the momentum flux term converts the area integral into a volume integral,

$$\int_S (\rho \mathbf{v}\cdot\mathbf{n})\,\mathbf{v}\,\mathrm{d}S = \int_V \nabla\cdot(\rho \mathbf{v}\mathbf{v})\,\mathrm{d}V \tag{4.18}$$

Thus, the momentum equation becomes,

$$\int_V \frac{\partial \rho \mathbf{v}}{\partial t}\,\mathrm{d}V = -\int_V \nabla\cdot(\rho \mathbf{v}\mathbf{v})\,\mathrm{d}V + \int_V \rho \mathbf{f}\,\mathrm{d}V - \int_V \nabla p\,\mathrm{d}V + \mathbf{F}_{\mathbf{visc}} \tag{4.19}$$

Defining $\boldsymbol{\mathcal{F}}_{\mathbf{visc}}$,

$$\mathbf{F}_{\mathbf{visc}} = \int_V \boldsymbol{\mathcal{F}}_{\mathbf{visc}}\,\mathrm{d}V \tag{4.20}$$

We can now write,

$$\int_V \left[\frac{\partial \rho \mathbf{v}}{\partial t} + \nabla\cdot(\rho \mathbf{v}\mathbf{v}) - \rho \mathbf{f} + \nabla p - \boldsymbol{\mathcal{F}}_{\mathbf{visc}}\right]\mathrm{d}V = 0 \tag{4.21}$$

As discussed in Sect. 4.4, this integral is independent of V. Thus, the general differential form of the momentum equation is,

$$\frac{\partial \rho \mathbf{v}}{\partial t} + \nabla\cdot(\rho \mathbf{v}\mathbf{v}) = -\nabla p + \rho \mathbf{f} + \boldsymbol{\mathcal{F}}_{\mathbf{visc}} \tag{4.22}$$

This equation can be written for each cartesian coordinate. For example, for the x coordinate direction, Eq. (4.22) becomes,

$$\frac{\partial \rho u}{\partial t} + \nabla\cdot(\rho u \mathbf{v}) = -\frac{\partial p}{\partial x} + \rho f_x + (\mathcal{F}_{\mathrm{x}})_{\mathrm{visc}} \tag{4.23}$$

where f_x is the x component of $\mathbf{f}$ and $(\mathcal{F}_{\mathrm{x}})_{\mathrm{visc}}$ is the x component of $\boldsymbol{\mathcal{F}}_{\mathbf{visc}}$.

4.6 Energy Equation

The energy equation is how we express the conservation of energy in fluid mechanics, also known as the first law of thermodynamics,

$$\delta q + \delta w = \mathrm{d}e \tag{4.24}$$

where δq is an incremental amount of heat added by the surroundings, δw is an incremental amount of work done by the surroundings and de is the change in internal energy of the system caused by these increments in heat and work.

Consider the control volume shown in Fig. 4.2 which we will again use for our analysis. As we are interested in the flow of fluid through the control volume, the energy equation will be derived by considering the rate of energy passing through the control volume V. We start by considering the first law of thermodynamics (Eq. 4.24) as a rate process and write it as a power balance,

$$\begin{aligned}&\text{Rate of heat added to control volume } V + \text{Rate of work done on control volume } V\\ &\quad = \text{Rate of total energy change of control volume, } V\end{aligned} \tag{4.25}$$

We can express the rate of heat addition to the control volume V by,

$$\text{Rate of heat addition} = \int_V \dot{q}\rho\, \mathrm{d}V \tag{4.26}$$

where $\dot{q}$ is the rate of heat addition per unit mass from the surroundings to the control volume. Additional viscous heating can occur by fluid shear stress, we define this as the general viscous heating term $\dot{Q}_{\text{visc}}$.

Work can be done on the control volume by the pressure acting on the surface, the body force acting on the volume or by viscous forces generated by shear stress at the surface. To develop mathematical expressions for these terms, we use the fact that work done on a particle can be defined as the force acting on the particle multiplied by the distance it moves under that force. Thus it follows that the rate of work done on a particle is the force multiplied by its velocity. For a control volume, we are interested in the rate of energy added or subtracted to the volume by work. These rates or energy *fluxes* will be developed below.

The rate of work done on the control volume by pressure can be written,

$$\text{Rate of work done on control volume by pressure} = -\int_S (p\mathbf{n}\, \mathrm{d}S)\cdot \mathbf{v} \tag{4.27}$$

This expression (Eq. 4.27) can be considered the sum of pressure forces acting on elemental areas (dS) multiplied by the velocity ($\mathbf{v}$) of fluid passing through each of these areas.

Unlike the pressure force, the body force per unit mass **f** acts on each part of the control volume. Thus, we determine the rate of work done by the body force as the sum of contributions to each elemental volume dV,

$$\text{Rate of work done on control volume by the body force} = -\int_V (\rho \mathbf{f}\, dV) \cdot \mathbf{v} \tag{4.28}$$

Additional work can be done by the viscous forces if there is fluid shear. We denote this viscous work as $\dot{W}_{\text{visc}}$.

The total energy per unit mass of a moving fluid is the sum of its internal energy e and its kinetic energy per unit mass $\frac{1}{2}U^2$,

$$\text{Total energy per unit mass} = e + \frac{U^2}{2} \tag{4.29}$$

where U is the mean local flow velocity magnitude,

$$U = |\mathbf{v}| = \sqrt{u^2 + v^2 + w^2} \tag{4.30}$$

As mass flows across the control surface, it transports its total energy with it. The rate of this total energy flow can be determined by summing the individual flows of energy per unit mass across each elemental are $\mathrm{d}S$ by the mass flow through $\mathrm{d}S$. We write this as,

$$\text{Rate of total energy flow across control surface} = \int_S (\rho \mathbf{v} \cdot \mathbf{n}\, \mathrm{d}S)\left(e + \frac{U^2}{2}\right) \tag{4.31}$$

We now will develop an expression for the time rate of change of total energy in the control volume. First, we determine the total energy in the control volume,

$$\text{Total energy in control volume} = \int_V \rho \left(e + \frac{U^2}{2}\right) \mathrm{d}V \tag{4.32}$$

The time rate of change is hence,

$$\text{Time rate of change of total energy in control volume} = \frac{\partial}{\partial t}\int_V \rho \left(e + \frac{U^2}{2}\right) \mathrm{d}V \tag{4.33}$$

We can now re-write the power balance shown in Eq. (4.25). First, each term will be explicitly written,

$$\text{Rate of heat added to control volume } V = \int_V \dot{q}\rho\, \mathrm{d}V + \dot{Q}_{\text{visc}} \tag{4.34}$$

$$\text{Rate of work done on control volume } V = -\int_S (p\mathbf{n}\,\mathrm{d}S)\cdot\mathbf{v} - \int_V (\rho\mathbf{f}\,dV)\cdot\mathbf{v} + \dot{W}_{\mathrm{visc}} \tag{4.35}$$

$$\begin{aligned}&\text{Rate of total energy change of control volume, } V\\ &= \int_S (\rho\mathbf{v}\cdot\mathbf{n}\,\mathrm{d}S)\left(e+\frac{U^2}{2}\right) + \frac{\partial}{\partial t}\int_V \rho\left(e+\frac{U^2}{2}\right)\mathrm{d}V\end{aligned} \tag{4.36}$$

Complete the power balance by equating the terms in Eq. (4.25),

$$\begin{aligned}&\int_V \dot{q}\rho\,\mathrm{d}V + \dot{Q}_{\mathrm{visc}} - \int_S (p\mathbf{n}\,\mathrm{d}S)\cdot\mathbf{v} - \int_V (\rho\mathbf{f}\,dV)\cdot\mathbf{v} + \dot{W}_{\mathrm{visc}}\\ &\quad = \int_S (\rho\mathbf{v}\cdot\mathbf{n}\,\mathrm{d}S)\left(e+\frac{U^2}{2}\right) + \frac{\partial}{\partial t}\int_V \rho\left(e+\frac{U^2}{2}\right)\mathrm{d}V\end{aligned} \tag{4.37}$$

Applying the divergence theorem to convert the surface integrals to volume integrals (as shown for the continuity and momentum equations earlier), we can obtain the differential form of the energy equation,

$$\begin{aligned}\frac{\partial}{\partial t}\left[\rho\left(e+\frac{U^2}{2}\right)\right] + \nabla\cdot\left[\rho\left(e+\frac{U^2}{2}\right)\mathbf{v}\right] &= \rho\dot{q} - \nabla\cdot(p\mathbf{v}) + \rho(\mathbf{v}\cdot\mathbf{v})\\ &\quad + \dot{\mathcal{Q}}_{\mathrm{visc}} + \dot{\mathcal{W}}_{\mathrm{visc}}\end{aligned} \tag{4.38}$$

where $\dot{\mathcal{Q}}_{\mathrm{visc}}$ and $\dot{\mathcal{W}}_{\mathrm{visc}}$ are the viscous heating and work terms in the partial differential form of the energy equation.

In order to close the fundamental equations, we require yet more equations. For an ideal, perfect gas, we can link internal energy and temperature,

$$e = c_v T \tag{4.39}$$

where c_v is the specific heat at constant temperature. The perfect gas equation of state can be used to link pressure, density and temperature,

$$p = \rho R T \tag{4.40}$$

where R is the specific gas constant.

4.7 Summary

The fundamental equations of fluid mechanics are as follows.

The continuity equation,

$$\frac{\partial \rho}{\partial t} + \nabla \cdot (\rho \mathbf{v}) = 0 \tag{4.8}$$

The momentum equation,

$$\frac{\partial \rho \mathbf{v}}{\partial t} + \nabla \cdot (\rho \mathbf{v}\mathbf{v}) = -\nabla p + \rho \mathbf{f} + \mathcal{F}_{\mathbf{visc}} \tag{4.22}$$

The energy equation,

$$\frac{\partial}{\partial t}\left[\rho\left(e + \frac{U^2}{2}\right)\right] + \nabla \cdot \left[\rho\left(e + \frac{U^2}{2}\right)\mathbf{v}\right] = \rho \dot{q} - \nabla \cdot (p\mathbf{v}) + \rho(\mathbf{v} \cdot \mathbf{v}) + \dot{\mathcal{Q}}_{\text{visc}} + \dot{\mathcal{W}}_{\text{visc}} \tag{4.38}$$

With closure using the thermodynamic relationship,

$$e = c_v T \tag{4.39}$$

and the perfect gas equation of state,

$$p = \rho R T \tag{4.40}$$

Chapter 5
Acoustics

Abstract This chapter reviews the basic concepts and physical processes of sound generation and propagation. We begin by discussing the nature of sound and the behaviour of sound waves. We examine how to quantify and analyse a sound signal using sound pressure level evaluation and frequency analysis. The acoustic wave equation that describes the propagation of acoustic disturbances in a quiescent fluid is derived along with the propagation properties of simple plane and spherical waves. We introduce the concept of sound intensity that describes the flow of acoustic energy in a sound field and the calculation of sound power. We conclude by deriving the sound radiation characteristics of common idealised sources including volume displacement, dipole and quadrupole sources.

5.1 Introduction

Acoustics is concerned with the production, propagation and effects of sound. Sound is a wave phenomenon that propagates through an elastic medium via pressure variations within it. Sound can be produced by two fundamental mechanisms: (1) a vibrating or pulsating object or (2) pressure fluctuations induced by turbulence and unsteady flows, referred to as flow-induced or aerodynamic sound. Sound waves can propagate through any fluid (liquid or gas) or solid but air is the most common medium through which sound travels. This chapter focuses mainly on sound in atmospheric air at audible frequencies.

This chapter presents an overview of fundamental aspects of acoustics related to the production, propagation and quantification of sound. Many of the definitions and theoretical formulations presented in this chapter are standardised and can be found in technical standard documentation (e.g. [4–7]). Those interested in more in-depth information on acoustics are referred to specialised texts [1, 2].

C. Doolan and D. Moreau, *Flow Noise*,
https://doi.org/10.1007/978-981-19-2484-2_5

5.2 Sound Waves

Sound propagates in the form of a longitudinal wave involving an alternating series of compressions and rarefactions of the fluid particles, as illustrated in Fig. 5.1. Compressions and rarefactions are high and low pressure regions in a fluid, respectively, and the distance between successive compressions or rarefactions is defined as the acoustic wavelength denoted by λ. The intersection point on the pressure axis in Fig. 5.1 represents the local mean quiescent pressure (normally the ambient atmospheric pressure), p_0. When a sound wave propagates in air, the oscillations in pressure are above and below the ambient atmospheric pressure which has a nominal value of 1.013×10^5 Pa (Pascals or Nm^{-2}) at sea level. The atmospheric pressure can vary by $\pm 5\%$ due to meteorological conditions and it decreases with increasing altitude. Everyday sounds produce variations in pressure that are rapid and extremely small compared with the ambient pressure value.

A distinguishing feature of a sound wave is that it propagates rapidly through a fluid medium at a speed, referred to as the speed of sound, that depends on the type of fluid and its local properties. Typical values of the speed of sound, c_0, are 343 and 1500 m/s in dry air at 20 °C and seawater at 13 °C, respectively.

A relationship exists between the wavelength of sound and frequency, f, according to

$$\lambda = \frac{c_0}{f} \tag{5.1}$$

Frequency is the rate at which the sound source produces complete cycles of compressions and rarefactions. The sensation of frequency is commonly referred to as the pitch of a sound. High frequencies result in high pitched or treble sounds while low frequencies are associated with low pitched or bass sounds. A healthy, young person can hear sounds with frequencies from 20 Hz to 20 kHz. The sound of human speech is typically in the range 300 Hz to 3 kHz.

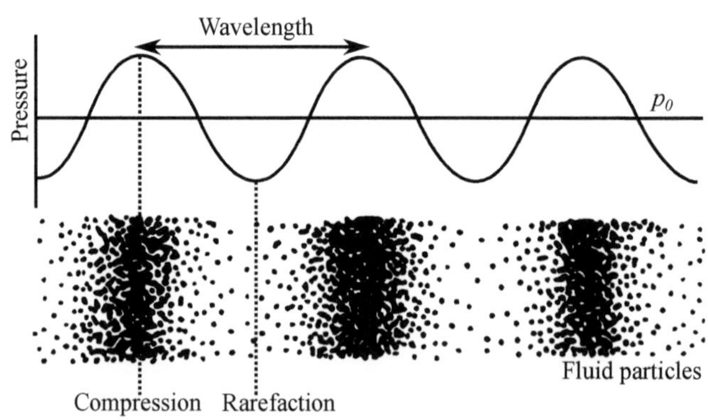

Fig. 5.1 Sound wave in air

5.3 Quantifying a Sound Signal

The most complete measurement of sound is its pressure-time history measured in pascals (Pa). The two most significant properties of a sound pressure signal are its amplitude and frequency content.

The amplitude of different sounds can vary by a significant amount. For example, the sound pressure of a jet aircraft at take-off (20 Pa) is 6 orders of magnitude greater than the lowest sound detectable by the human ear (20 μPa). Given the very large range of sound pressures, a logarithmic measure of pressure referred to as Sound Pressure Level (or SPL) and expressed in decibels (dB), is used to compress the range of typical noise events from 0 to 140 dB. This scale also mimics the approximately logarithmic response of the human auditory system.

Sound pressure level is a measure of the Root-Mean-Square (RMS) pressure, p_{RMS}, of sound according to

$$L_p = 10 \log_{10}(p_{RMS}^2 / p_{ref}^2) \tag{5.2}$$

$$= 20 \log_{10}(p_{RMS} / p_{ref}) \tag{5.3}$$

where p_{ref} is a reference pressure. For air, the standard reference pressure is 20 μPa which corresponds to the threshold of hearing for a young person with normal hearing over the frequency range of 1–4 kHz. For water, the standard reference pressure is 1 μPa. It is good practice to always state the reference pressure when reporting SPL. To illustrate the relationship between perceived loudness and sound pressure level, Table 5.1 states the sound pressure level (dB re 20 μPa) of various common sound sources and Table 5.2 reports the subjective assessment of changes in SPL.

RMS pressure, p_{RMS}, is used to quantify sound pressure that varies with time. To calculate the RMS pressure, the values of a pressure-time history are squared, summed and then averaged over some arbitrary time interval, T. The square root is then taken of this average value to produce an RMS value according to

Table 5.1 Sound pressure level (dB re 20 μPa) of common sound sources

Description of sound source or environment	SPL
Moon launch at 100 m, artillery fire	140
Jet aircraft take-off at 100 m, live rock music	120
Motorcycle, farm tractor, handheld drill	100
Busy highway at 20 m, garbage disposal, shouting	80
Conversation in a restaurant or office, normal speech	60
Quiet urban ambient sound, library	40
Whisper, rustling leaves	20
Threshold of hearing for normal young people	0

Table 5.2 Subjective assessment of changes in SPL (dB re 20 μPa)

Change in SPL	Change in apparent loudness
3	Just perceptible
5	Clearly noticeable
10	Half or twice as loud
20	Much quieter or louder

$$p_{RMS} = \sqrt{\frac{1}{T}\int_0^T p^2(t)dt} \tag{5.4}$$

In practice, sound pressure is measured with a microphone that obtains instantaneous pressure samples at specific time intervals. To calculate the RMS pressure in this case, each individual sample is first squared and then summed. The result is divided by the total number of samples, N, and the square root is taken to give the RMS pressure

$$p_{RMS} = \sqrt{\frac{1}{N}\sum_{n=1}^{N} p^2(n)} \tag{5.5}$$

One method of determining the frequency content of a sound pressure signal is to use a Fourier transform as described in Sect. 3.6. To permit comparison of acoustic measurements, standardised frequency bands may be used as specified by the International Standards Organisation. The widest band employed in frequency analysis is the octave band. Each octave band is described by its centre frequency and the upper frequency limit of the octave band is double the lower frequency limit. If more detailed information about the frequency content is needed, each octave can be divided into three equal bands (on a logarithmic scale) to produce standardised one-third octave bands. Table 5.3 states the frequency bands for octave and one-third octave band frequency analysis. The band centre frequencies and the lower and upper band limits, denoted f_C, f_L and f_U, respectively, can be calculated according to

$$f_C = 10^{n/10} \tag{5.6}$$

$$f_L = 10^{(n-0.5)/10} \tag{5.7}$$

$$f_U = 10^{(n+0.5)/10} \tag{5.8}$$

where n is the frequency band number. Table 5.3 shows that ten octaves span the audible frequency range (20 Hz to 20 kHz). To evaluate the octave or one-third octave band acoustic spectrum, a band limited filter is applied to the pressure-time history and the sound pressure level in each band is then calculated.

Table 5.3 Octave and one-third octave frequency bands (Hz) and A-weighting corrections (dB) [3]

Band number	Octave band centre frequency	One-third octave band centre frequency	Lower band limit	Upper band limit	Octave band A-weighting correction	One-third octave band A-weighting correction
14		25	22	28		−44.7
15	31.5	31.5	28	35	−39.4	−39.4
16		40	35	44		−34.6
17		50	44	57		−30.2
18	63	63	57	71	−26.2	−26.2
19		80	71	88		−22.5
20		100	88	113		−19.1
21	125	125	113	141	−16.1	−16.1
22		160	141	176		−13.4
23		200	176	225		−10.9
24	250	250	225	283	−8.6	−8.6
25		315	283	353		−6.6
26		400	353	440		−4.8
27	500	500	440	565	−3.2	−3.2
28		630	565	707		−1.9
29		800	707	880		−0.8
30	1,000	1,000	880	1,130	0.0	0.0
31		1,250	1,130	1,414		0.6
32		1,600	1,414	1,760		1.0
33	2,000	2,000	1,760	2,250	1.2	1.2
34		2,500	2,250	2,825		1.3
35		3,150	2,825	3,530		1.2
36	4,000	4,000	3,530	4,400	1.0	1.0
37		5,000	4,400	5,650		0.5
38		6,300	5,650	7,070		−0.1
39	8,000	8,000	7,070	8,800	−1.1	−1.1
40		10,000	8,800	11,300		−2.5
41		12,500	11,300	14,140		−4.3
42	16,000	16,000	14,140	17,600	−6.6	−6.6
43		20,000	17,600	22,500		−9.3

The human auditory system perceives different spectral components of an acoustic signal with different loudness. It is most sensitive to sound in the frequency range between 1 and 4 kHz. To mimic this response, a weighting network, referred to as the A-weighting curve, can be applied to the acoustic spectrum to attenuate frequency components below 1 kHz and above 4 kHz. The A-weighting curve is the most common weighting network used in environmental noise measurements and its values are given in Table 5.3. Other weighting networks include B and C-weighting curves which were designed to mimic the human auditory response to high sound levels of 55–85 dB and above 85 dB, respectively. A D-weighting curve was also proposed specifically for aircraft noise measurements but it is rarely used.

5.4 Acoustic Wave Equation

Having discussed pressure fluctuations at a point, we now consider sound generation and propagation. The acoustic wave equation is the fundamental equation describing the propagation of acoustic disturbances in a quiescent fluid. This equation is derived from the conservation equations for fluids (introduced in Chap. 4) and uses the three quantities of pressure, density and particle velocity to describe acoustic fluctuations.

We consider the situation where there is no significant flow and the acoustic waves are the only source of pressure and velocity fluctuations. The total pressure, p is given by the sum of the ambient pressure, p_0 and a time varying perturbation (or the acoustic pressure), p' according to $p = p_0 + p'$ where $p' \ll p_0$. Similarly, when a fluid is compressed or rarefied, its density will vary from its ambient value, ρ_0, such that the total density $\rho = \rho_0 + \rho'$, where ρ' is the density fluctuation and $\rho' \ll \rho_0$. We assume that the ambient pressure, p_0 and density, ρ_0 are independent of time and uniform throughout the medium. As there is no mean flow, $\mathbf{v}$ is the velocity fluctuations due to acoustic waves only.

Sound waves disturb the ambient pressure only a tiny bit so the variables associated with sound (p', ρ' and $\mathbf{v}$) are small quantities of first order. This is termed the small-signal approximation and is used to linearize the conservation equations. We first consider the continuity equation in Eq. 4.8 and replace ρ with $\rho_0 + \rho'$. Assuming that the product of two small quantities (e.g. $\rho'\mathbf{v}$) is negligible and that ρ_0 is a constant, the linearized continuity equation is

$$\frac{\partial \rho'}{\partial t} + \nabla \cdot (\rho_0 \mathbf{v}) + \nabla \cdot (\rho' \mathbf{v}) = \frac{\partial \rho'}{\partial t} + \rho_0 \nabla \cdot \mathbf{v} = 0 \tag{5.9}$$

Following the same procedure, the momentum equation in Eq. 4.22 can be linearized to give

$$\rho_0 \frac{\partial \mathbf{v}}{\partial t} + \nabla p' = 0 \tag{5.10}$$

If we differentiate the linearized continuity equation (Eq. 5.9) with respect to time and the linearized momentum equation (Eq. 5.10) with respect to distance (x) and subtract, we obtain the linear wave equation

$$\frac{\partial^2 \rho'}{\partial t^2} - \nabla^2 p' = 0 \tag{5.11}$$

Pressure is often employed as the primary field variable in acoustics and therefore we need to eliminate the density perturbation in Eq. (5.11). The perfect gas equation of state in Eq. (4.40) relates the state variables of pressure, density and temperature to one another. However, we assume that acoustic disturbances propagate with negligible thermal energy transfer between adjacent fluid elements and the entropy of the fluid, denoted by s, remains constant. Therefore, pressure is a direct function of density only according to

$$p = p(\rho) \tag{5.12}$$

This is the general form of the isentropic equation of state. We may express Eq. (5.12) as a Taylor series expansion as a function of density according to

$$p = p_0 + A\frac{\rho - \rho_0}{\rho_0} + \frac{B}{2!}\left(\frac{\rho - \rho_0}{\rho_0}\right)^2 + \frac{C}{3!}\left(\frac{\rho - \rho_0}{\rho_0}\right)^3 + \cdots \tag{5.13}$$

where A, B, C, ... are coefficients. If the fluctuations are small, the value of $(\rho - \rho_0)/\rho_0$ is also small and the most important coefficient is therefore A. To evaluate A, we require the speed of sound, c_0 defined as

$$c_0^2 \equiv \left(\frac{\partial p}{\partial \rho}\right)_{s=const} \tag{5.14}$$

$$= \frac{dp}{d\rho} \text{ for an isentropic process} \tag{5.15}$$

Differentiating Eq. (5.13) and evaluating the limit $\rho \rightarrow \rho_0$ gives

$$A = \rho_0 c_0^2 \tag{5.16}$$

Technically, the quantity c_0 should be called the small-signal sound speed as it is only true in the limit of $\rho \rightarrow \rho_0$. Using Eqs. (5.13) and (5.16), we can now express the isentropic equation of state in the form

$$p' = c_0^2 \rho' \left[1 + \frac{B}{2!A}\frac{\rho'}{\rho_0} + \frac{C}{3!A}\left(\frac{\rho'}{\rho_0}\right)^2 \dots\right] \tag{5.17}$$

Linearizing this equation defines the linear relationship between pressure and density fluctuations as

$$p' = c_0^2 \rho' \tag{5.18}$$

Eq. (5.18) is used to replace the density perturbation in Eq. (5.11) to give the acoustic wave equation

$$\frac{1}{c_0^2}\frac{\partial^2 p'}{\partial t^2} - \nabla^2 p' = 0 \tag{5.19}$$

This is a second order linear partial differential equation that relates the way in which acoustic pressure fluctuations behave with respect to space and time. The next step is to find solutions that satisfy the acoustic wave equation.

5.5 Simple Solutions to the Acoustic Wave Equation

While sound propagation is complicated in practice, it can be useful to examine the propagation properties of simple plane and spherical waves in one dimension. The characteristic property of a plane wave, shown in Fig. 5.2, is that each acoustic variable has constant amplitude and phase in planes perpendicular to the direction of wave propagation. An example is the acoustic pressure waves traveling along a thin tube or duct. For a plane wave that propagates along the x-direction, the wave equation reduces to

$$\frac{1}{c_0^2}\frac{\partial^2 p'}{\partial t^2} - \frac{\partial^2 p'}{\partial x^2} = 0 \tag{5.20}$$

The most general solution of Eq. (5.20) is known as d'Alembert's solution given by

$$p'(x, t) = f(t - x/c_0) + g(t + x/c_0) \tag{5.21}$$

where f and g are arbitrary functions determined by boundary and initial conditions. In the x, t (space, time) plane, functions $f(t - x/c_0)$ and $g(t - x/c_0)$ describe pressure disturbances propagating in the positive and negative x-directions, respectively. For a wave traveling in the positive x-direction, Eq. (5.21) shows that the disturbance $f(t)$ generated at $x = 0$ is reproduced at a distance x along the duct at a time x/c_0 later. The same applies for the disturbance traveling in the negative x-direction.

Fig. 5.2 Plane wave fronts

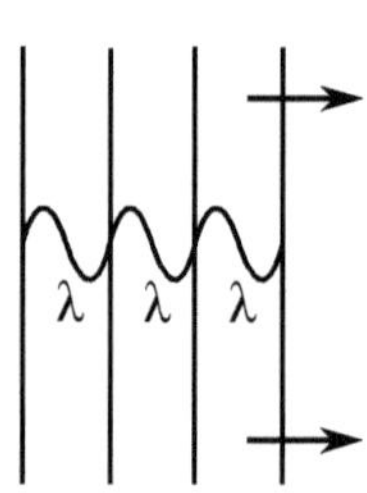

Of particular interest is the time-harmonic form of the solution. Harmonic signals consist of sinusoidal variations at a single frequency and are often referred to as pure tones. In time-harmonic form, function f (and similarly function g) can be written as

$$f(t - x/c_0) = A\cos(\omega t - kx + \phi) \tag{5.22}$$

where $\omega = 2\pi f$ is the angular frequency, $k = \omega/c_0$ is the acoustic wavenumber and terms A and ϕ are amplitude and phase constants, respectively. For a harmonic disturbance, the time period of the wave oscillation is $T = 1/f = 2\pi/\omega$ and the spatial distribution of the sound field also oscillates with length scale given by the acoustic wavelength $\lambda = 2\pi/k$.

Using complex notation with a time-dependence of $e^{i\omega t}$, function f can also be written as

$$f(t - x/c_0) = \text{Re}\left[\hat{A}e^{i(\omega t - kx)}\right] \tag{5.23}$$

where ˆ indicates complex amplitude and $\hat{A} = Ae^{i\phi}$. The complex notation is defined such that the physical quantity to be measured is described by the real component. This form of the harmonic solution to the wave equation is often employed in mathematical manipulations as it allows waves of different phases to be easily added together and simplifies the operations of integration and differentiation.

A general solution to the wave equation also exists for spherical waves illustrated in Fig. 5.3. In this case, the waves are a function of radial distance r from the centre of the co-ordinate system. An example is the propagation of sound waves from a small source in free space with no nearby boundaries. Reducing the Laplacian operator of Eq. (5.19) so that it is a function of radial distance, the wave equation may be rearranged and put in the form

$$\frac{1}{c_0^2}\frac{\partial^2 p'}{\partial t^2} - \frac{1}{r}\frac{\partial^2}{\partial r^2}(rp') = 0 \tag{5.24}$$

When the source is located at the origin, the solution to Eq. (5.24) is given by

$$rp' = f(t - r/c_0) + g(t + r/c_0) \tag{5.25}$$

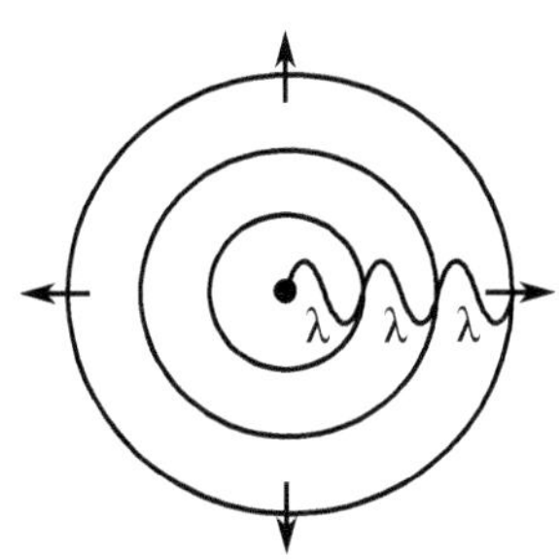

Fig. 5.3 Spherical wave fronts

or

$$p'(r,t) = \frac{f(t - r/c_0)}{r} + \frac{g(t + r/c_0)}{r} \tag{5.26}$$

The first term of Eq. (5.26) represents a wave propagating spherically outwards from the origin in the radial direction. Importantly, the wave amplitude reduces according to $1/r$ showing that as the distance r increases, the wave amplitude linearly decreases. The corresponding sound pressure level decreases by 6 dB per doubling of distance. The second term of Eq. (5.26) represents an incoming, spherically converging wave whose amplitude increases as it converges towards the origin.

The converging wave is rare in practice and in most cases, we are only interested in the outwardly propagating wave with solution

$$p'(r,t) = \frac{f(t - r/c_0)}{r} \tag{5.27}$$

It is useful to consider the complex form of the harmonic solution for the acoustic pressure of the spherical wave given by

$$p'(r,t) = \text{Re}\left[\frac{\hat{A}e^{i(\omega t - kr)}}{r}\right] \tag{5.28}$$

We can also define the spatially dependent complex pressure amplitude $\hat{p}(r)$ as

$$\hat{p}(r) = \frac{\hat{A}e^{-ikr}}{r} \tag{5.29}$$

It is common to ignore the time dependence and employ the solution in Eq. (5.29) in acoustic analysis as we can recover the actual pressure fluctuation from

$$p'(r,t) = \text{Re}\left[\hat{p}(r)e^{i\omega t}\right] \tag{5.30}$$

5.6 Sound Power and Intensity

When an acoustic wave propagates through a fluid, energy is transmitted from one part of the fluid to another. Instantaneous sound intensity, or the rate at which work is done by an element of fluid on a adjacent element, per unit area, is

$$\mathbf{I}_i = p'\mathbf{v} \tag{5.31}$$

The general expression for sound intensity, I, in a specified direction is the time averaged rate at which energy flows through a unit area, where the normal to the area points in the specified direction. This is given by

$$\mathbf{I} = \frac{1}{T}\int_0^T p'\mathbf{v}dt \tag{5.32}$$

where sound intensity has units of watts per meter squared (Wm^{-2}). The value of measurement time T depends on the type of acoustic signal being measured. For periodic signals, T is the period. For non-periodic signals including random noise, T must be a sufficiently long time.

A simple relationship exists between the acoustic pressure and particle velocity in a plane wave that can be used to derive a special expression for its sound intensity. For an arbitrary forward traveling wave, f, in Eq. (5.21), the linearised equation of momentum (Eq. 5.10) becomes

$$\rho_0 \frac{\partial v}{\partial t} = \frac{1}{c_0} f'(t - x/c_0) \tag{5.33}$$

where the prime sign in f' is used to indicate differentiation of the function by its argument ($\mathrm{d}f(w)/\mathrm{d}w = f'(w)$). From this equation it can be seen that $\partial p'/\partial t = \rho_0 c_0 \partial v/\partial t$ and therefore

$$p' = \rho_0 c_0 v \tag{5.34}$$

This equation shows that the acoustic pressure and particle velocity are related by the quantity $\rho_0 c_0$, termed the acoustic impedance. The acoustic impedance has a value of 414 Ns/m^3 at 20 °C and one atmosphere. From Eq. (5.34), it follows that the general sound intensity expression for plane and traveling waves is

$$I = \frac{p_{RMS}^2}{\rho_0 c_0} \tag{5.35}$$

In practice, Eq. (5.34) rarely describes the relationship between the pressure and particle velocity of a sound wave. Instead, it is useful to derive a general expression for the sound intensity associated with harmonic pressure and particle velocity fluctuations. For harmonic pressure fluctuations $p'(x,t) = \mathrm{Re}\left[\hat{p}(x)e^{i\omega t}\right]$ where $\hat{p}(x)$ is the complex pressure amplitude, we can write

$$\mathrm{Re}\left[\hat{p}(x)e^{i\omega t}\right] = \mathrm{Re}\left[(p_R + ip_I)(\cos\omega t + i\sin\omega t)\right] = p_R\cos\omega t - p_I\sin\omega t \tag{5.36}$$

where $\hat{p}(x)$ is a complex number $p_R + ip_I$. In the same manner, for harmonic velocity fluctuations $v(x,t) = \mathrm{Re}\left[\hat{v}(x)e^{i\omega t}\right]$, we can write

$$\mathrm{Re}\left[\hat{v}(x)e^{i\omega t}\right] = v_R\cos\omega t - v_I\sin\omega t \tag{5.37}$$

where $\hat{v}(x)$ is the complex velocity amplitude written as a complex number $v_R + iv_I$. Substituting Eqs. (5.36) and (5.37) into Eq. (5.32) and evaluating the integral (using trigonometric relations) gives the following expression for sound intensity

$$\mathbf{I} = \frac{1}{2}(p_R v_R + p_I v_I) = \frac{1}{2}\mathrm{Re}\left[\hat{p}^*(x)\hat{v}(x)\right] = \frac{1}{2}\mathrm{Re}\left[\hat{p}(x)\hat{v}^*(x)\right] \tag{5.38}$$

where * denotes the complex conjugate. This equation can be used to determine the sound intensity for harmonic fluctuations in pressure and velocity.

Sound intensity can be used to determine sound power, W. The power passing through a surface S is the integral of the intensity over that surface according to

$$W = \int_S \mathbf{I} \cdot \mathbf{n} dS \tag{5.39}$$

where $\mathbf{n}$ is the unit vector normal to the surface of area S and sound power has the units of watts (W). Most often, a surface encompassing a sphere or spherical section is used to calculate sound power. However, other shaped surfaces may be employed depending on the particular configuration under consideration. For a sound source producing spherical waves, calculation of the sound power using a spherical surface with radius r, leads to the following expression

$$W = 4\pi r^2 I \tag{5.40}$$

In this case the source has been considered to radiate uniformly in all directions. Meanwhile, for plane waves in a one-dimensional duct in which the intensity is uniform over the cross-sectional area S, the power is simply

$$W = IS \tag{5.41}$$

Similar to the range of acoustic pressures encountered in practice, the power output of different sounds can vary by a significant amount. For example, the power output of a whispered voice and jet engine are 10^{-9} W and 10^4 W, respectively. To compress this vast range, a logarithmic measure of sound power is used termed sound power level (or PWL). The sound power level is expressed in decibels (dB) and is given by

$$L_W = 10\log_{10}(W/W_{ref}) \tag{5.42}$$

where W_{ref} is the reference power of 10^{-12} W. Similarly, sound intensity can be expressed as the sound intensity level (or IL) in decibels (dB) according to

$$L_I = 10\log_{10}(I/I_{ref}) \tag{5.43}$$

where the reference intensity in air is 10^{-12} Wm^{-2}. For the special case of traveling plane and spherical waves, IL and SPL are related by

$$L_I = 10\log_{10}\left(\frac{p_{RMS}^2}{\rho_0 c_0 I_{ref}}\frac{p_{ref}^2}{p_{ref}^2}\right) \tag{5.44}$$

$$= 20\log_{10}\frac{p_{RMS}}{p_{ref}} + 10\log_{10}\frac{p_{ref}^2}{\rho_0 c_0 I_{ref}} \tag{5.45}$$

The first term of Eq. (5.45) is SPL leading to

$$L_I = L_P - 0.16 \approx L_P \tag{5.46}$$

It follows that for practical purposes, SPL and IL are numerically equivalent for progressive waves in air.

5.7 Sound Sources

5.7.1 *The Pulsating Sphere*

Let us first consider a pulsating sphere that contracts and expands with time. The sphere has radius a and harmonic normal surface velocity of amplitude v_0 as shown in Fig. 5.4. In this case, the appropriate solution that matches the boundary condition is Eq. (5.29). To determine the unknown constant $\hat{A}$, we need to match the particle velocity of the sound wave in the radial direction to the surface velocity of the pulsating sphere.

For a harmonic wave, we define the radial component of the complex particle velocity as $\hat{v}_r(r)$ where the actual radial particle velocity fluctuation is $v_r(r,t) = \text{Re}\left[\hat{v}(r)e^{i\omega t}\right]$. Using Eq. (5.10), the acoustic momentum equation for a harmonic wave is

$$i\omega\rho_0\hat{v}_r(r) = \nabla\hat{p}(r) \tag{5.47}$$

This equation relates the complex particle velocity in the radial direction to the gradient of the acoustic pressure in that direction. Differentiating Eq. (5.29) with

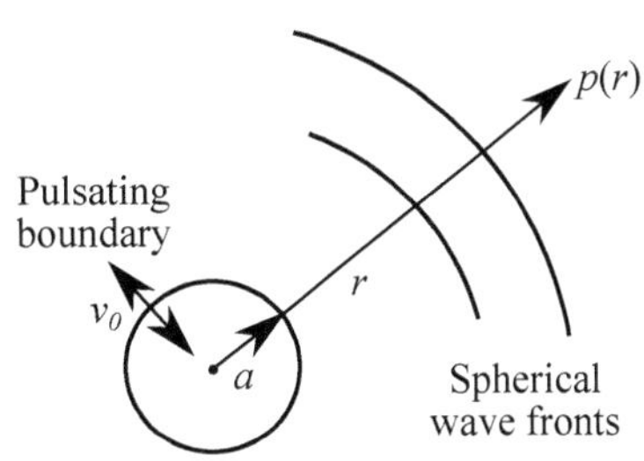

Fig. 5.4 Pulsating sphere of radius a

respect to r and substituting the result into Eq. (5.47) gives the following expression for the complex radial particle velocity

$$\hat{v}_r(r) = \frac{\hat{A}(1 + ikr)}{i\omega\rho_0 r^2} e^{-ikr} \tag{5.48}$$

For the pulsating sphere with surface velocity $\left[\hat{v}_r\right]_{r=a} = v_0$, Eqs. (5.29) and (5.48) can be used to write the following expression for the complex pressure at $r = a$

$$\hat{p}(a) = \frac{\hat{A}e^{-ika}}{a} = \rho_0 c_0 \left[\frac{ika}{1 + ika}\right] v_0 \tag{5.49}$$

Equation (5.49) can be used to determine unknown constant $\hat{A}$ as a function of v_0. Substituting this into Eq. (5.29) gives the complete solution for the complex pressure at a radial distance r as

$$\hat{p}(r) = \rho_0 c_0 a \left[\frac{ika}{1 + ika}\right] v_0 \frac{e^{-ik(r-a)}}{r} \tag{5.50}$$

Equation (5.50) shows that the sound field is determined by the boundary conditions only. As a time-harmonic pressure variation was assumed, initial conditions are not required in the pressure calculation.

We can define the source strength, Q, as the rate of change of the volume of the sphere or the *volume velocity*. It is the product of the surface area of the sphere, $S = 4\pi a^2$, and its radial velocity. In this case, the volume velocity is a harmonic quantity with amplitude given by

$$Q = 4\pi a^2 v_0 \tag{5.51}$$

so Eq. (5.50) becomes

$$\hat{p}(r) = \rho_0 c_0 \left[\frac{ik}{1 + ika}\right] \frac{Qe^{-ik(r-a)}}{4\pi r} \tag{5.52}$$

If the sphere is very small such that $ka \ll 1$ and $a/r \ll 1$, then Eq. (5.50) approximates to

$$\hat{p}(r) = \frac{i\omega\rho_0 Qe^{-ikr}}{4\pi r} \tag{5.53}$$

This equation shows that the acoustic pressure is directly proportional to the rate of change of the volume of the sphere caused by its surface displacement. The sound field radiates uniformly in all directions and is a function of the radial distance, r, from the centre of the sphere only. This source is commonly referred to as a volume displacement source or an acoustic *monopole*. Many practical sources of mechanical and aerodynamic noise can be approximated as monopoles. Examples include a loudspeaker mounted inside a rigid, closed cabinet at low frequencies or the sound radiation from the exhaust of a reciprocating engine.

It is useful to note that in Eq. (5.53), the term $i\omega\rho_0 Q$ is the harmonic monopole source strength and the expression for the pressure per unit source strength is known as the free-field Green's function, G, given by

$$G = \frac{e^{-ikr}}{4\pi r} \tag{5.54}$$

We now consider the sound power radiated by a monopole. An expression for the sound intensity associated with spherical wave propagation is first required. From Eq. (5.38) for harmonic fluctuations, the sound intensity in the radial direction, r, can be expressed in terms of the complex pressure and radial particle velocity according to

$$I_r = \frac{1}{2}\text{Re}\left[\hat{p}(r)\hat{v}_r^*(r)\right] \tag{5.55}$$

Evaluating this equation (and using Eqs. (5.48) and (5.53)), gives the radial sound intensity

$$I_r = \frac{|\hat{p}(r)|^2}{2\rho_0 c_0}\text{Re}\left[\frac{1+ikr}{ikr}\right] = \frac{|\hat{p}(r)|^2}{2\rho_0 c_0} \tag{5.56}$$

as a function of the modulus squared of the complex pressure only. According to Eq. (5.39), the total power output is found by integrating the sound intensity over a surface which encloses the source. For a monopole source, we select a spherical surface that lies in the far-field ($kr \gg 1$) and the unit normal vector, n, points outwards in the radial direction. The sound power of a monopole as defined in Eq. (5.53) is therefore given by

$$W = \frac{(\omega\rho_0 Q)^2}{2\rho_0 c_0 (4\pi r)^2}\int_S dS \tag{5.57}$$

As the surface area of a sphere of radius r is $4\pi r^2$, the sound power of a monopole is

$$W = \frac{(\omega\rho_0 Q)^2}{8\pi\rho_0 c_0} \tag{5.58}$$

5.7.2 Multipole Sources

Section 5.7.1 introduced the fundamental volume displacement source referred to as an acoustic monopole. Attention now turns to multipole sources that are composed of multiple simple sources located in close proximity to one another.

A *dipole* consists of two monopoles of equal strength oscillating with opposite phase, as illustrated in Fig. 5.5. The monopoles are separated by a distance d that is very small compared with an acoustic wavelength ($kd \ll 1$). In this case, there is no

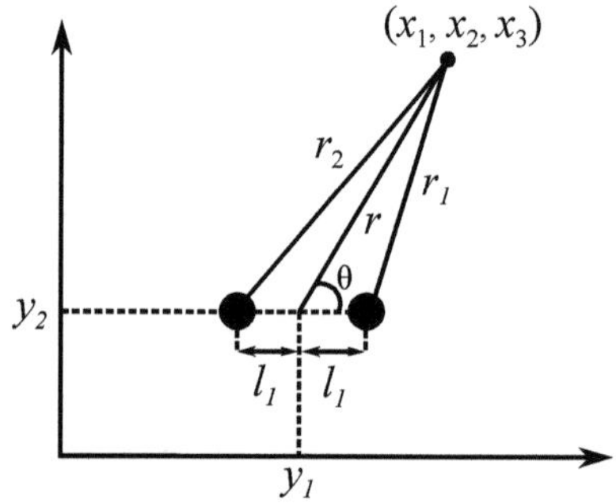

Fig. 5.5 The geometry of a dipole source consisting of two sources of equal strength and opposite phase

resultant mass flow through a spherical surface surrounding the two sources but there is a momentum flux resultant. This is equivalent to a net fluctuating force applied to the fluid.

To obtain the dipole sound field, we assume that the source is centred at $\mathbf{y} = (y_1, y_2, y_3)$ and the observer location is $\mathbf{x} = (x_1, x_2, x_3)$, as shown in Fig. 5.5. The two monopole sources have equal and opposite strength denoted Q and $-Q$ and their separation distance is $d = 2l_1$ where l_1 is small. The radial distances, r_1 and r_2, between the two monopole sources and the measurement location are given by

$$r_1 = \sqrt{(x_1 - (y_1 + l_1))^2 + (x_2 - y_2)^2 + (x_3 - y_3)^2} \tag{5.59}$$

and

$$r_2 = \sqrt{(x_1 - (y_1 - l_1))^2 + (x_2 - y_2)^2 + (x_3 - y_3)^2} \tag{5.60}$$

Using Eq. (5.53), the complex pressure produced by the dipole source can be written as

$$\hat{p}(\mathbf{x}) = i\omega\rho_0 Q \left[\frac{e^{-ikr_1}}{4\pi r_1} - \frac{e^{-ikr_2}}{4\pi r_2} \right] \tag{5.61}$$

We can express the terms inside the square brackets of Eq. (5.61) as functions $f(y_1 \pm l_1)$ such that by using a Taylor series of the form

$$f(y_1 + l_1) = f(y_1) + l_1 \frac{\partial}{\partial y_1} f(y_1) + \frac{1}{2!} l_1^2 \frac{\partial^2}{\partial y_1^2} f(y_1) + \dots \tag{5.62}$$

we can write

$$\frac{e^{-ikr_1}}{4\pi r_1} = \frac{e^{-ikr}}{4\pi r} + l_1 \frac{\partial}{\partial y_1} \frac{e^{-ikr}}{4\pi r} + \frac{1}{2!} l_1^2 \frac{\partial^2}{\partial y_1^2} \frac{e^{-ikr}}{4\pi r} + \dots \tag{5.63}$$

A similar expression may also be written for $\frac{e^{-ikr_2}}{4\pi r_2}$ leading to terms that are a function of the radial distance r from the centre of the source to the observer. In the limit

$l_1 \to 0$, only the leading-order terms in the series contribute and the complex pressure is therefore given by

$$\hat{p}(r) = i\omega\rho_0 Q 2l_1 \frac{\partial}{\partial y_1}\left[\frac{e^{-ikr}}{4\pi r}\right] \tag{5.64}$$

To evaluate this equation we use $d = 2l_1$ and the chain rule $\frac{\partial}{\partial y_1} = \frac{\partial r}{\partial y_1}\frac{\partial}{\partial r}$ where

$$\frac{\partial r}{\partial y_1} = \frac{(y_1 - x_1)}{r} = -\cos\theta \tag{5.65}$$

from Eq. (5.59) and Fig. (5.5),

$$\frac{\partial}{\partial y_1}\left[\frac{e^{-ikr}}{4\pi r}\right] = \frac{ike^{ikr}}{4\pi r}\left[1 - \frac{i}{kr}\right] \tag{5.66}$$

and θ is the angle of the vector from the centre of the dipole $\mathbf{y}$ to the observer location $\mathbf{x}$ relative to the horizontal axis (see Fig. 5.5). The pressure radiated by the dipole source is therefore given by

$$\hat{p}(r) = -\frac{\omega\rho_0 Q d e^{-ikr}}{4\pi r}(k\cos\theta)\left[1 + \frac{1}{ikr}\right] \tag{5.67}$$

From this equation, the near-field pressure radiated close to the dipole is characterized by the term $1/ikr$. In the far-field where $kr \to \infty$, the expression for the radiated pressure becomes

$$\hat{p}(r) = (ikd\cos\theta)\,\hat{p}_m \tag{5.68}$$

where $\hat{p}_m$ is the far-field pressure radiated by a monopole given in Eq. (5.53). Equation (5.68) shows the maximum amplitude of the dipole sound field is ikd times the amplitude of a monopole field at the same distance. As $kd \ll 1$, the dipole source is an inefficient radiator. Another unique feature of dipole radiation is its figure-eight directivity pattern due to the $\cos\theta$ term in Eq. (5.67) as shown in Fig. 5.6. The sound radiated by the dipole is strongest in the direction of the dipole axis which is oriented along the line that joins the two monopole sources. At 90° to the dipole axis, the two sources cancel producing a null or zero.

A *quadrupole* is comprised of two dipoles placed next to each another. The dipoles can be arranged in back-to-back (see Fig. 5.7) or side-to-side configuration (see Fig. 5.8) to produce a longitudinal or lateral quadrupole, respectively. A quadrupole source produces no net unbalanced force but a net stress is applied to the fluid. Quadrupole sources are particularly important in aeroacoustics as they can be used to explain the noise produced by turbulence in the mixing region of a free jet that exhausts into a quiescent atmosphere.

First, consider a longitudinal quadrupole located at $\mathbf{y}$ consisting of two sources with strength Q at $(y_1 + l_1, y_2, y_3)$ and $(y_1 - l_1, y_2, y_3)$ and a source with strength

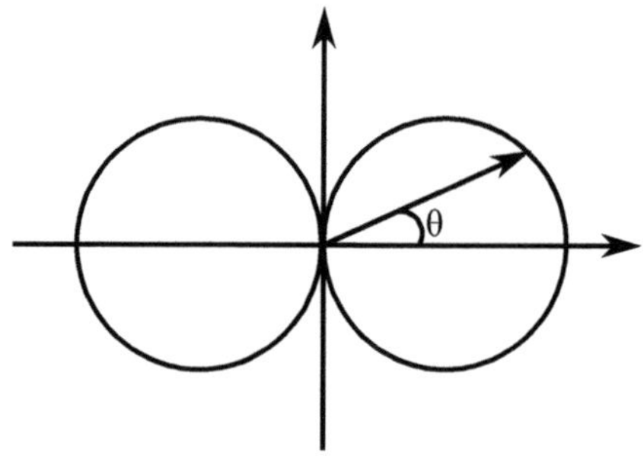

Fig. 5.6 The radiation pattern of a dipole source indicating the angular variation in sound pressure amplitude. The axis of the source is horizontal and the directivity is axi-symmetric about the horizontal axis

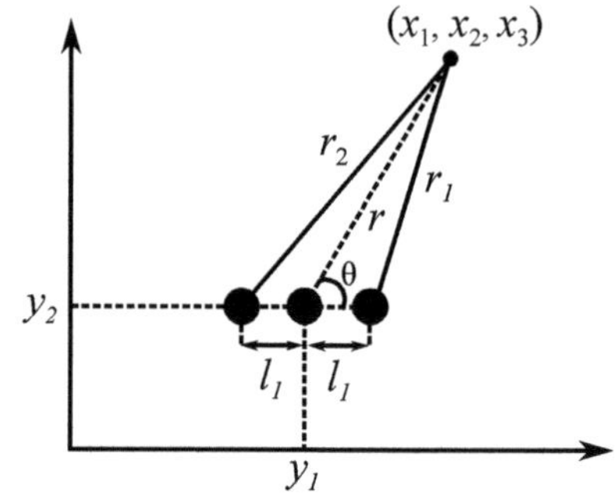

Fig. 5.7 The geometry of a longitudinal quadrupole source

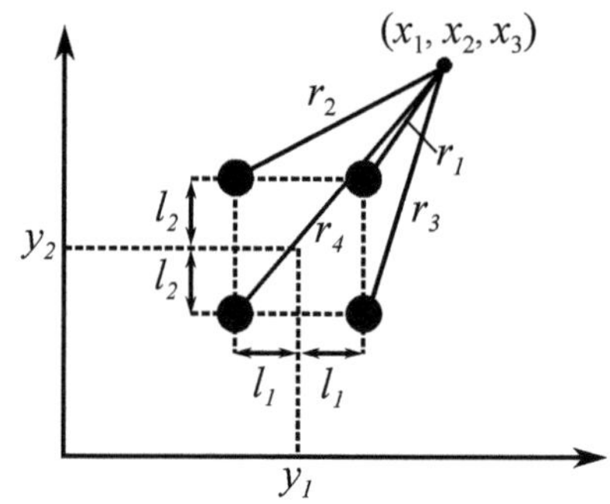

Fig. 5.8 The geometry of a lateral quadrupole source

$-2Q$ at **y**, as shown in Fig. 5.7. Using Eq. (5.53), the complex pressure produced by the quadrupole source can be written as

$$\hat{p} = i\omega\rho_0 Q\left[\frac{e^{-ikr_1}}{4\pi r_1} - \frac{2e^{-ikr}}{4\pi r} + \frac{e^{-ikr_2}}{4\pi r_2}\right] \tag{5.69}$$

Applying the expansion procedure outlined in Eq. (5.63) leads to

$$\hat{p} = i\omega\rho_0 Q d^2 \frac{\partial^2}{\partial y_1^2}\left[\frac{e^{-ikr}}{4\pi r}\right] \tag{5.70}$$

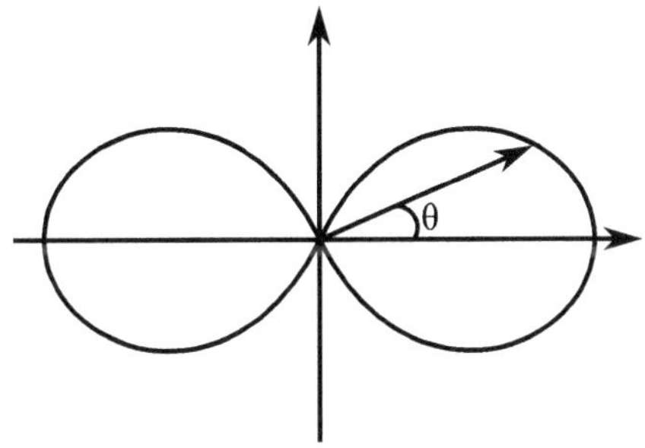

Fig. 5.9 The radiation pattern of a longitudinal quadrupole source indicating the angular variation in sound pressure amplitude

where the distance between individual sources is $d = l_1$. Evaluating the second derivative in Eq. (5.70) produces the following expression for the pressure radiated by a longitudinal quadrupole

$$\hat{p} = -\frac{\omega\rho_0 Q e^{-ikr}}{4\pi r}(kd)^2\left[\cos^2\theta\left(1+\frac{3}{ikr}+\frac{3}{(ikr)^2}\right)-\frac{1}{ikr}-\frac{1}{(ikr)^2}\right] \quad (5.71)$$

where θ is the angle of the vector from the centre of the quadrupole $\mathbf{y}$ to the observer location $\mathbf{x}$ relative to the horizontal axis. In the near-field when kr is small, quadrupole sound radiation is dominated by the $(1/(kr)^2)$ terms. In the far-field where $kr \to \infty$, the expression for the radiated pressure becomes

$$\hat{p}(r) = (kd\cos\theta)^2\,\hat{p}_m \quad (5.72)$$

This equation shows that the maximum amplitude of the quadrupole sound field is $(kd)^2$ times the amplitude of a monopole field at the same distance. The directivity of longitudinal quadrupole sound radiation is shown in Fig. 5.9 and is dependent on the $\cos^2\theta$ term in Eq. (5.72).

Next, consider the case of a lateral quadrupole in Fig. 5.8. The quadruple centred at $\mathbf{y}$ consists of two sources with strength Q at $(y_1 - l_1, y_2 - l_2, y_3)$ and $(y_1 + l_1, y_2 + l_2, y_3)$ and two sources with strength $-Q$ at $(y_1 + l_1, y_2 - l_2, y_3)$ and $(y_1 - l_1, y_2 + l_2, y_3)$. Using Eq. (5.53), the pressure field produced by the quadrupole is

$$\hat{p} = i\omega\rho_0 Q\left[\frac{e^{-ikr_1}}{4\pi r_1}-\frac{e^{-ikr_2}}{4\pi r_2}-\frac{e^{-ikr_3}}{4\pi r_3}+\frac{e^{-ikr_4}}{4\pi r_4}\right] \quad (5.73)$$

which can be reduced to

$$\hat{p} = i\omega\rho_0 Q d^2\frac{\partial^2}{\partial y_1\partial y_2}\left[\frac{e^{-ikr}}{4\pi r}\right] \quad (5.74)$$

where the distance between sources is $d = 2l_1 = 2l_2$. Evaluating the partial derivatives gives the following expression for the pressure radiated by a lateral quadrupole

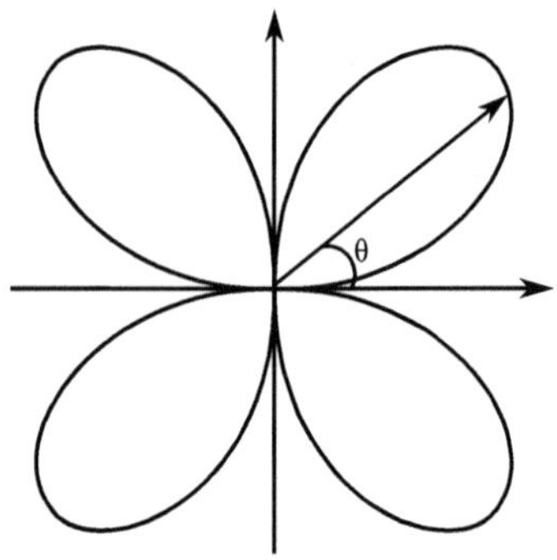

Fig. 5.10 The radiation pattern of a lateral quadrupole source indicating the angular variation in sound pressure amplitude

$$\hat{p} = -\frac{\omega \rho_0 Q e^{-ikr}}{4\pi r}(kd)^2 \left[\cos\theta \sin\theta \left(1 + \frac{3}{ikr} + \frac{3}{(ikr)^2}\right)\right] \tag{5.75}$$

The far-field sound pressure radiated by a lateral quadrupole is smaller than that of a monopole by a factor of $(kd)^2$ at the same distance and its directivity is dependent on $\cos\theta \sin\theta = 1/2 \sin 2\theta$ as shown in Fig. 5.10.

The sound intensity and power output of a multipole source can be estimated by following the same procedure as outlined in Sect. 5.7.1. Considering sound radiation in the acoustic far-field, the sound power of a dipole is given by

$$W = \frac{(\omega \rho_0 Q)^2 (kd)^2}{24\pi \rho_0 c_0} \tag{5.76}$$

In the case of a quadrupole, the sound power output is

$$W = \frac{(\omega \rho_0 Q)^2 (kd)^4}{10\pi \rho_0 c_0} \tag{5.77}$$

The sound power radiated by a dipole is smaller than that of a single monopole source by a factor proportional to $(kd)^2$ (where we have assumed $kd \ll 1$) due to cancellation of the sound field produced by its two individual monopole components. The sound power radiated by a quadrupole is smaller than that of a dipole and a monopole by a factor proportional to $(kd)^2$ and $(kd)^4$, respectively. For any source distribution that features a combination of monopole, dipole and quadrupole sources, the sound power output will be dominated by the monopole sources.

5.8 Summary

This chapter has presented an introduction to acoustics. After the fundamentals are reviewed, quantification of a sound signal in terms of its amplitude and frequency content is discussed. Following this, sound wave propagation is described using the acoustic wave equation. The concepts of sound intensity and sound power are presented before the chapter concludes with a description of sound radiation from idealised monopole, dipole and quadrupole sources.

References

1. Bies D, Hansen C (2009) Engineering noise control: theory and practice, 4th edn. Spon Press
2. Blackstock DT (2000) Fundamentals of physical acoustics. Wiley-Interscience
3. EC 60651 (2001) Sound level meters. Standard, International Electrotechnical Commission
4. IEC 60050–801 (1994) International electrotechnical vocabulary—chapter 801: acoustics and electroacoustics. Standard, International Electrotechnical Commission
5. ISO 12001 (1996) Acoustics—noise emitted by machinery and equipment—rules for the drafting and presentation of a noise test code. Standard, International Organization of Standards
6. ISO 3744 (2010) Acoustics—determination of sound power levels and sound energy levels of noise sources using sound pressure—engineering methods for an essentially free field over a reflecting plane. Standard, International Organization of Standards
7. ISO 9614–1 (1993) Acoustics—determination of sound power levels of noise sources using sound intensity—part 1: measurement at discrete points. Standard, International Organization of Standards

Chapter 6
Laminar and Turbulent Flow

Abstract This chapter will provide an overview of laminar and turbulent flows. Laminar flows are treated first, with canonical flows such as the flat plate boundary layer, plane mixing layer, free jet and wake flows described. Transition is covered, with the Orr-Sommerfeld equation used to explain instabilities in transitional boundary layer flows. Turbulent flows are described, with topics including statistical descriptions of turbulence, Reynolds averaging, spectral models, turbulent boundary layers and wall pressure spectral models.

6.1 Introduction

Flow-induced noise is created when there is unsteady flow. The state of a flow is controlled, in part, by the Reynolds number. Before we get to a formal definition, we can think of the Reynolds number as the ratio of the flow's inertia to its viscous force. Laminar flows are found when the Reynolds number is low; the case where viscous forces are much greater than the flow's inertia thus keeping the flow stable. When the Reynolds number increases, inertial forces become more important and the flow becomes unstable. When the Reynolds number is moderate, the flow is said to be transitional. This is when instabilities are beginning to form as the inertial forces are just starting to overcome the stabilising viscous forces. Increasing the Reynolds number further can push the flow into the turbulent regime where the flow is unstable and inertial forces dominate over the viscous ones.

Turbulent flows are found nearly everywhere. For the understanding and analysis of flow-induced noise, we are usually concerned with a sub-set of this infinite number of flow situations. These include jets, wakes, shear flows and boundary layers.

As we will see in later chapters, acoustic waves are generated by the unsteady motion of fluid. This unsteady motion creates forces on objects and stresses within the fluid which are sources of sound. Both sound and turbulence can be described as waves; however, there are critical differences. The first relates to dissipation. Sound is essentially non-dissipative over reasonably short distances whereas turbulence is dissipative. Dissipation is a key part of energy management within turbulent flows.

C. Doolan and D. Moreau, *Flow Noise*,
https://doi.org/10.1007/978-981-19-2484-2_6

Another key difference is wave speed. Sound waves move at the speed of sound, whereas turbulence eddies move at their local convection wave speed (often assumed to be the local mean velocity). For low Mach number flows, this means the wavelength of sound generated by turbulence is much larger than the convective wavelength. For higher Mach number flows, theses scales can become comparable. In any case, the generation of sound by unsteady flow is multi-scale problem, where the length scales of the source are much smaller than the length scales of the sound produced.

This chapter will provide an overview of laminar and turbulent flows as a foundation to understanding flow induced noise later in the book. Laminar flows are treated first, with canonical flows such as the flat plate boundary layer, plane mixing layer, free jet and wake flows described. Transition is covered, with the Orr-Sommerfeld equation used to explain instabilities in transitional boundary layer flows. Turbulent flows are then described, with topics including statistical descriptions of turbulence, Reynolds averaging, spectral models, turbulent boundary layers and wall pressure spectral models.

6.2 Laminar Flow

A fundamental dimensionless parameter of importance to this chapter is the Reynolds number,

$$Re_x = \frac{U_\infty x}{\nu} \tag{6.1}$$

where Re_x is the Reynolds number evaluated using length-scale x, U_∞ is the freestream (or reference) velocity and ν is the kinematic viscosity of the fluid. The Reynolds number can be interpreted as the ratio of inertial forces to the viscous forces with in the flow.

Below a certain value of the Reynolds number (The indifference Reynolds number [32]), viscous forces dominate the flow field and the flow has certain characteristic that we label as *laminar flow*. The flow is called laminar because it appears the be layered [32]. Each layer moves over the top of each other with little exchange of mass between them. In fact, in laminar flow, fluid particles are only exchanged via a process of molecular diffusion, which is very slow.

6.2.1 Laminar Boundary Layer on a Flat Plate with Zero Pressure Gradient

Figure 6.1 illustrates the basic ideas concerning laminar flow in a boundary layer. Uniform flow with freestream velocity U_∞ encounters a flat plate. The viscous stress induced by the plate surface (imposed by the no-slip boundary condition) causes

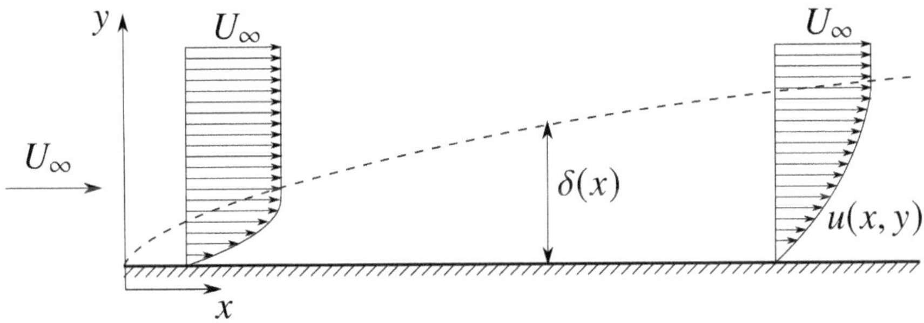

Fig. 6.1 Illustration of a laminar boundary layer on a zero-pressure-gradient flat plate

the velocity field to change. In fact, this can be interpreted as infinitesimal layers (or *laminae*) of fluid sliding over each other with different velocity. The streamwise velocity is now a two-dimensional field represented by $u(x, y)$. The streamwise (x) positions where the velocity field equals 99% of U_∞ is known as the boundary layer height or thickness. The boundary layer height is a function of x only and is represented by $\delta(x)$ and is shown as a dashed line in Fig. 6.1.

To understand the zero-pressure-gradient laminar boundary more completely, we should consider the two-dimensional, steady and incompressible continuity and momentum equations,

$$\frac{\partial u}{\partial x} + \frac{\partial v}{\partial y} = 0 \tag{6.2}$$

$$u\frac{\partial u}{\partial x} + v\frac{\partial v}{\partial y} = \nu\frac{\partial^2 u}{\partial y^2} \tag{6.3}$$

Note that because of the zero pressure gradient, $\frac{\partial p}{\partial x} = 0$, leaving only the viscous stress term on the right hand side of the momentum equation. Consistent with the rest of the book, u is streamwise velocity, v is vertical velocity, x is the streamwise coordinate and y is the vertical coordinate.

The boundary conditions are, formally:

$$y = 0, \qquad u = 0,\ v = 0 \tag{6.4}$$

$$y \to \infty, \qquad u = U_\infty \tag{6.5}$$

Following Schlicting [32], we assume self-similar solutions for the boundary layer velocity. In fact, a similarity variable can be defined,

$$\eta = y\sqrt{\frac{U_\infty}{2\nu x}} \tag{6.6}$$

This can be used to define a stream function,

$$\psi = \sqrt{2\nu x U_\infty} f(\eta) \tag{6.7}$$

where $f(\eta)$ is a dimensionless stream function. In this form, the velocity components are,

$$u = \frac{\partial \psi}{\partial y} = \frac{\partial \psi}{\partial \eta}\frac{\partial \eta}{\partial y} = U_\infty f'(\eta) \tag{6.8}$$

$$v = -\frac{\partial \psi}{\partial x} = \frac{\partial \psi}{\partial \eta}\frac{\partial \eta}{\partial x} = \sqrt{\frac{\nu U_\infty}{2x}}(\eta f' - f) \tag{6.9}$$

Substituting Eqs. (6.8) and (6.9) into the momentum equation, Eq. (6.3), gives the *Blasius equation* which is the ordinary differential equation for the stream function,

$$f''' + ff'' = 0 \tag{6.10}$$

with boundary conditions,

$$\begin{aligned} \eta = 0, &\quad f = 0,\ f' = 0 \\ \eta \to \infty, &\quad f' = 1 \end{aligned}$$

The Blasius equation is commonly used to solve for laminar boundary layer problems. It should be noted that more general forms can be obtained to include pressure gradient [32]. The Blasius equation is solved numerically and important solutions are summarised below.

The skin-friction coefficient (c_f) for a laminar flat-plate boundary layer is found by first deducing the wall shear stress (τ),

$$\tau_w(x) = \mu\left(\frac{\partial u}{\partial y}\right) = \mu U_\infty \sqrt{\frac{U_\infty}{2\nu x}} f''_w = 0.332\mu\sqrt{\frac{U_\infty}{\nu x}} \tag{6.11}$$

Here, $f''_w = 0.4696$ was found numerically [32].

Subsequently, the skin-friction coefficient is,

$$c_f(x) = \frac{\tau_w(x)}{\frac{1}{2}\rho_\infty U_\infty^2} = \frac{0.664}{\sqrt{Re_x}} \tag{6.12}$$

The *boundary-layer thickness*, defined at the value of $y = \delta_{99}$ when $u = 0.99U_\infty$, is similarly found to be,

$$\frac{\delta_{99}}{x} \approx \frac{5.0}{\sqrt{Re_x}} \tag{6.13}$$

The *displacement thickness* (δ^*) is an integral dimension of the boundary layer. Physically, it represents the amount the wall must be moved to compensate for the mass 'lost' by the velocity deficit in the boundary layer, compared with a uniform flow over a slip-wall boundary condition. It is defined as,

$$\delta^* = \int_0^\infty \left(1 - \frac{u}{U_\infty}\right) \mathrm{d}y \tag{6.14}$$

Using the Blasius velocity profile for a laminar boundary layer (Eq. 6.13), the displacement thickness for a flat plate becomes,

$$\frac{\delta^*}{x} = \frac{1.7208}{\sqrt{Re_x}} \tag{6.15}$$

The *momentum thickness* (θ) is a length scale related to the loss of momentum due to the viscous shear in the boundary layer. It can be thought of as the distance the wall must be moved into the flow in order to achieve a uniform flow with same momentum as the flow including the boundary layer. It is defined as,

$$\theta = \int_0^\infty \frac{u}{U_\infty}\left(1 - \frac{u}{U_\infty}\right) \mathrm{d}y \tag{6.16}$$

For a laminar, flat-plate boundary layer (using Eq. 6.13), the momentum thickness is,

$$\frac{\theta}{x} = \frac{0.664}{\sqrt{Re_x}} \tag{6.17}$$

It is also useful to know the ratios of displacement and momentum thicknesses to the boundary-layer height. These are listed below for the laminar flat-plate boundary layer,

$$\frac{\delta^*}{\delta_{99}} = 0.34 \tag{6.18}$$

$$\frac{\theta}{\delta_{99}} = 0.13 \tag{6.19}$$

6.2.2 Plane Mixing Layer

When two fluids with different but parallel velocities encounter each other, they form a mixing layer (or shear layer), as shown in the schematic Fig. 6.2. The simple situation, shown in Fig. 6.2, is the case where two parallel streams are initially separated by a splitter plate. This can represent practical applications such as the lip of a nozzle,

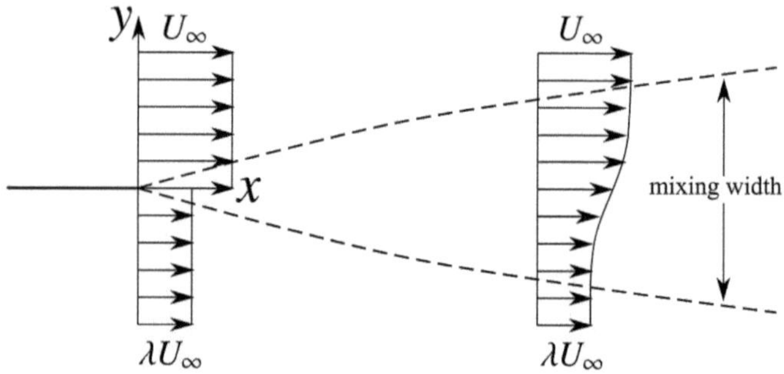

Fig. 6.2 Illustration of a laminar plane mixing layer

the flow over the leading edge of a cavity or the trailing edge of an airfoil. Other situations where shear layers form are in areas of separated flow. Here, a boundary layer (another form of shear layer) separates from the wall, forming a mixing layer separating two fluids with different velocities. This often occurs on airfoils near or beyond the stall angle. Airfoils operating at low Reynolds number (around the transition Reynolds number) also often contain regions of separated flow near their surfaces. The instability of these laminar shear layers can lead to tonal noise generation.

We introduce the laminar mixing layer here as a canonical flow. Two parallel flows exist on either side of a splitter plate, the upper velocity is U_∞ and the lower velocity is λU_∞ (Fig. 6.2). At the end of splitter plate, the two flows, each with viscosity ν, interact. In this case, the Reynolds number is low enough to ensure laminar flow in the mixing layer. Thus, each layer of fluid in the mixing layer slides over each other with only a little mass transport which is controlled by molecular diffusion.

We assume zero pressure gradient along the mixing layer and the y component of velocity is small compared with the x component. Thus, the Blasius equation can be used to describe the flowfield as a similarity solution, but with different boundary conditions [31, 32],

$$f''' + ff'' = 0 \tag{6.20}$$

with boundary conditions,

$$\eta \to \infty, \qquad f' = 1$$
$$\eta \to -\infty, \qquad f' = \lambda$$

This equation must be solved numerically [22]. In some situations, the velocity profile can be approximated by a hyperbolic tan profile [3, 4, 24]. Blake [3] suggests the following hyperbolic tan velocity profile for the case where $\lambda = 0$,

$$u(y) = \frac{U_\infty}{2}(1 + \tanh y) \tag{6.21}$$

Blake also suggests a representative mixing-width length scale (δ) for such a profile is the position where the curvature of the velocity profile is a maximum; that is, where

$$\frac{d^3u(y=\delta)}{dy^3} = 0 \tag{6.22}$$

6.2.3 Plane Free Jet

The laminar plane (two-dimensional) free jet is illustrated in Fig. 6.3. Like the boundary layer and mixing layer, the free jet exists in a zero-pressure-gradient flow. Thus, it is tempting to apply the Blasius equation directly to this problem as well; however, some care must be taken. The free jet used in our discussion here is formed by uniform flow of velocity U_j issuing into a parallel co-flow of velocity U_∞. A special case where $U_\infty = 0$ is often encountered in practice and will be analysed below.

First the flowfield can be characterised to occur in separate zones. Close to the jet outlet (far-left of Fig. 6.3), the velocity is initially discontinuous across the edge of the free jet where the flow changes instantaneously from U_j to U_∞. This sudden change in velocity causes a shear layer (mixing layer, Sect. 6.2.2) to form on either side of the jet. Within these shear layers is the *potential core*, the region where the flow is still parallel and uniform. As the shear layer grows, they eventually merge and close the potential core, after which a mixing layer is formed. The flowfield close to the jet exit and including the potential core is known as the near field. The Blasius equations are not valid in the near field region.

Schlictling [32] provides a self-similar solution to the laminar plane free jet in the far field (where $x \gg h$ where h is the jet exit height). This is done by modelling the jet as a potential source. Such a flow does not accurately model the flow in near field,

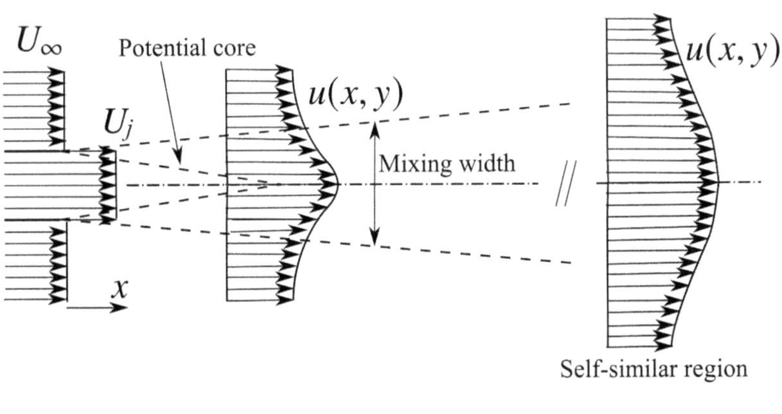

Fig. 6.3 Illustration of a laminar plane free jet

nor does it recreate the flow properly in the transition region between the near and far field; thus, solutions are limited to when the velocity profiles become self similar.

Taking the case where $U_\infty = 0$, a modified Blasius equation can be used to describe the self-similar region [31, 32],

$$f''' + ff'' + f'^2 = 0 \tag{6.23}$$

with boundary conditions,

$$\eta \to -\infty, \quad f' = 0$$
$$\eta = 0, \quad f = 0$$
$$\eta \to +\infty, \quad f' = 0$$

The total streamwise (x) momentum flux is estimated as [31],

$$J = \int_{-\infty}^{\infty} \rho u^2 \, \mathrm{d}y \tag{6.24}$$

As there is no net-shear acting on the flow over $-\infty < y < \infty$, J must remain constant.

Schlicting [32] and Schetz [31] give the following solution to Eq. (6.23),

$$u(x, y) = \frac{2}{3} a^2 x^{-\frac{1}{3}} [1 - \tanh^2(a\eta(x, y))] \tag{6.25}$$

where

$$\eta(x, y) = \frac{1}{3\nu^{\frac{1}{2}}} \frac{y}{x^{\frac{2}{3}}} \tag{6.26}$$

and

$$a = \left(\frac{9J}{16\rho\nu^{\frac{1}{2}}} \right)^{\frac{1}{3}} \tag{6.27}$$

6.2.4 Wake Flow

Steady laminar flow over a cylinder creates a plane wake, which is illustrated in Fig. 6.4. Uniform inflow approaches a cylinder of diameter D from the left. As it interacts with the cylinder it forms a wake with a *velocity deficit* which is a maximum towards the centre. The region close to the cylinder is known as the near wake. Here the velocity profile ($u(x, y)$) is not self-similar, thus numerical solutions of the Navier Stokes and continuity equations are normally used here. In the far wake where $y/D \gg 1$, the velocity profile becomes self-similar and a Gaussian profile

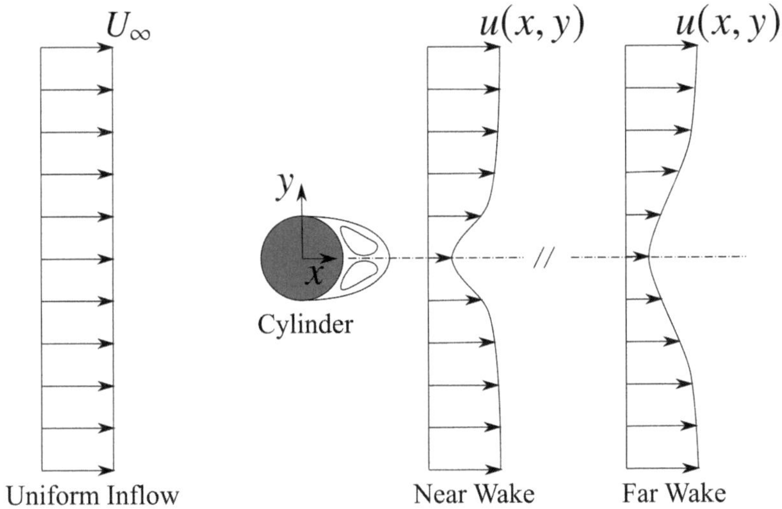

Fig. 6.4 Illustration of a laminar plane cylinder wake

can be used. Using a momentum balance in the far wake, Schetz [31] provides the following useful solution,

$$\frac{U_\infty - u(x, y)}{U_\infty} = C_D \left(\frac{Re_D}{16\pi} \frac{D}{x} \right)^{1/2} \exp\left(-\frac{U_\infty y^2}{4\nu x} \right) \tag{6.28}$$

where C_D is the drag coefficient of the cylinder and Re_D is the Reynolds number based on cylinder diameter.

6.3 Transition

As the Reynolds number increases, a laminar flow will *transition* to a turbulent state where the initial steady flow becomes unstable and eventually forms a chaotic unsteady flow field. Figure 6.5 illustrates this process for a boundary layer formed by the interaction of a uniform flow with a flat plate. Initially, a laminar boundary layer is formed. As the fluid travels over the plate, disturbances (acoustic waves, small vibrations, imperfections in the plate surface) at certain wavenumbers may be amplified to create two-dimensional Tollmien-Schlichting (TS) waves [32]. These waves become unstable and form three-dimensional waves, which, in turn are replaces by turbulent spots. As the Reynolds number further increases, these turbulent spots increase in number and merge to form a fully-turbulent flowfield. Thus, transition occurs over time and space, beginning with the initial onset of TS waves at the indif-

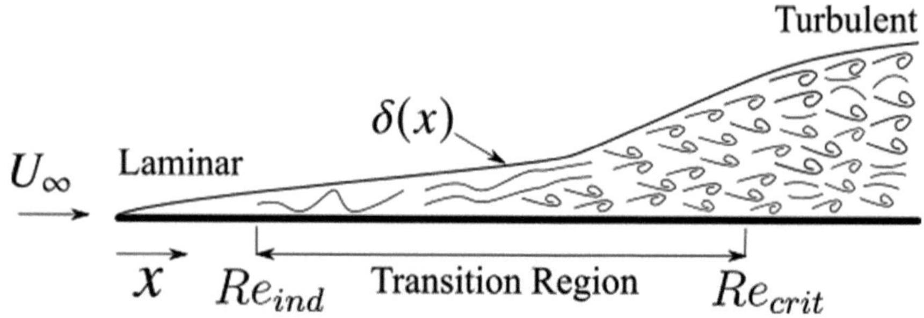

Fig. 6.5 Illustration of the transition process in a flat-plate boundary layer

ference Reynolds number, Re_{ind} and is complete at the critical Reynolds number, Re_{crit} [32].

The critical Reynolds number can vary depending on the flow conditions, disturbance levels and the leading-edge geometry of the flat plate. Is typically,

$$2.5 \times 10^5 \leq Re_{crit} \leq 10^6 \tag{6.29}$$

The onset of TS waves can be estimated using linear-stability theory. Here, we assume that two-dimensional TS waves are small disturbances imposed upon the mean two-dimensional laminar base flow [31],

$$u(x, y, t) = u_0(y) + u'(x, y, t) \tag{6.30}$$

$$v(x, y, t) = v'(x, y, t) \tag{6.31}$$

where we assume that the laminar base flow is function of y only and is represented by streamwise base velocity distribution $u_0(y)$. We assume that the disturbances are small, so that $u' \ll u_0$, hence the streamwise velcocity is steady ($u = u_0(y)$) and the only disturbance is $v'(x, y, t)$.

A stream function can be used to represent the perturbation [32],

$$\psi(x, y, t) = \hat{\phi}(y) \exp[i(\alpha x - \beta t)] \tag{6.32}$$

where $\hat{\phi}(y)$ is a complex amplitude (mode) function, α is the wavenumber of the disturbance and β is a complex quantity,

$$\beta = \beta_r + i\beta_i \tag{6.33}$$

where $\beta_r = \omega = 2\pi f$ is the frequency of the mode and β_i is an amplification factor. The wavenumber is $\alpha = 2\pi/\lambda$. Rewriting Eq. (6.32) we can show,

$$\psi(x, y, t) = \hat{\phi}(y) \exp\{i[\alpha x - (\beta_r + i\beta_i t)]\}$$
$$\psi(x, y, t) = \hat{\phi}(y) \exp(\beta_i t) \exp[i(\alpha x - \beta_r t)]$$
$$\psi(x, y, t) = \hat{\phi}(y) \exp(\beta_i t) \exp[i(\alpha x - \omega t)] \tag{6.34}$$

Thus, we can see that the value of β_i is linked to the instability of the waves: if $\beta_i < 0$, the waves are damped but if $\beta_i > 0$, the waves are amplified and indicates that the mode is unstable. Further, we can define the complex phase velocity,

$$c = \frac{\beta}{\alpha} = c_r + ic_i \tag{6.35}$$

where c_r is the velocity of the wave (in the x direction) and c_i is related to decay or amplification. The stream function can now be written,

$$\psi(x, y, t) = \hat{\phi}(y) \exp[i\alpha(x - ct)] \tag{6.36}$$

Using our understanding of stream functions, we can write expressions for the disturbance velocities,

$$u' = \frac{\partial \psi}{\partial y} = \frac{\partial \hat{\phi}}{\partial y} \exp[i\alpha(x - ct)] \tag{6.37}$$

$$v' = -\frac{\partial \psi}{\partial x} = -i\alpha\hat{\phi}(y) \exp[i\alpha(x - ct)] \tag{6.38}$$

Using these with the linearised, incompressible and unsteady Navier-Stokes equations, we can obtain the famous Orr-Sommerfeld equation [31, 32],

$$(u_0 - c)(\phi'' - \alpha^2\phi) - u_o''\phi = -\frac{i}{\alpha Re}(\phi'''' - 2\alpha^2\phi'' + \alpha^4\phi) \tag{6.39}$$

noting that the prime $'$ represents partial differentiation in the y direction. The Reynolds number is defined,

$$Re = \frac{U_e\delta^*}{\nu} \tag{6.40}$$

where U_e is the velocity external to the boundary layer (very close to U_∞ for the flat-plate laminar boundary layer).

The Orr-Sommerfeld equation is difficult to solve analytically and Eq. (6.39) is normally solved numerically [16]. Classically, linear velocity profiles and inviscid flow assumptions ($Re \to \infty$) were used in the past to obtain solutions and greater physical understanding. The most important result of these studies is the realisation that velocity profiles with points of inflection are unstable.

To understand this more clearly, consider a boundary layer subject to a pressure gradient. We know that a favourable pressure gradient (one that is decreasing with streamwise distance) has no point of inflection. In contrast, boundary layers in adverse (streamwise increasing) pressure gradients will form a point of inflection. This is known from boundary layer theory and our observations of the boundary layer separation process. Hence we expect that flows under adverse pressure gradient will transition more quickly than those under a favourable pressure gradient. This is indeed observed on airfoils at positive angle of attack. The upper or suction surface is more unstable then the lower or pressure surface, which have adverse and favourable pressure gradients respectively.

Moving back to the flat-plate boundary layer, we assume that this is a limiting (neutral) case and that there is a small inflection point at the surface for stability analysis purposes [32]. The lowest Reynolds number for which a neutral perturbation can exist is called the indifference Reynolds number (using Schlicting's nomenclature, see Sect. 6.5),

$$Re_{ind} = \frac{U_\infty \delta^*}{\nu} = 520 \tag{6.41}$$

which occurs for the following maximum wave speed and wave number,

$$\frac{c_r}{U_\infty} = 0.39 \quad \alpha\delta^* = 0.36 \tag{6.42}$$

Hence, the minimum wavelength of these TS waves is,

$$\lambda_{min} = \frac{2\pi\delta^*}{0.36} = 17.45\delta^* = 5.93\delta \tag{6.43}$$

using Eq. (6.18) to relate the boundary layer height to displacement height. Equation (6.43) shows that the wavelengths of TS waves can be large.

It is important to note that the onset of instability is only the beginning of the transition process. Transition is completed when the flow is completely turbulent and occurs at Re_{crit}. Equation (6.29) provides the range of values of Re_{crit} where transition is expected to be completed. Further, linear stability theory only applies at the beginning of the process, where the instabilities can be considered small and linearly superimposed on the laminar base flow. The processes between the onset of instability and full turbulence are non-linear and difficult to describe without numerical means or via extensive experimental observation.

A final note on transition. The gradual pathway of transition described above is not the only means of transition. An alternative pathway, known as *bypass* transition has also been observed [26]. This is a much earlier and sudden transition to turbulence that can't be explained by stability theory. It can occur when freestream disturbances are high (say 1% turbulence intensity [15, 32]) or when there are surface roughness effects, or other large-scale disturbances [26]. A turbulence 'trip', commonly used to ensure a turbulent boundary layer in small-scale wind tunnels, employs the bypass transition mechanism. Most university-scale wind tunnels have freestream turbulence levels in the range 0.1–1%, making transition measurements susceptable to bypass transition effects.

6.4 A Description of Turbulent Flow

As we have alluded to earlier in this chapter, turbulence is a chaotic or *random* process where kinetic energy is dissipated by viscous shear. This is in contrast to laminar flow, which is assumed steady and ordered into its 'laminae'. Even though turbulent flow is random, it has certain properties and behaviours.

Turbulence can be thought of as a kinetic energy dissipation process. For example, at high enough Reynolds numbers, the boundary layer on a plate will become turbulent due to instability in the flow (Sect. 6.3). Above the critical Reynolds number the flow is completely turbulent. Skin friction on the surface of the plate creates drag and velocity gradients that feed instabilities in the flow. These instabilities produce vortical flow features (eddies or coherent structures) whose kinetic energy must be dissipated. The process this occurs by governs turbulent flows and is called the energy cascade [29].

Figure 6.6 illustrates the turbulent energy cascade. It is assumed that the turbulence consists of eddies of various sizes (termed scales in the area of turbulence). The characteristic scale and velocity of each eddy is ℓ and $u(\ell)$ [28]. Consequently, each eddy has a 'turnover' timescale of approximately,

$$\tau(\ell) \equiv \frac{\ell}{u(\ell)} \tag{6.44}$$

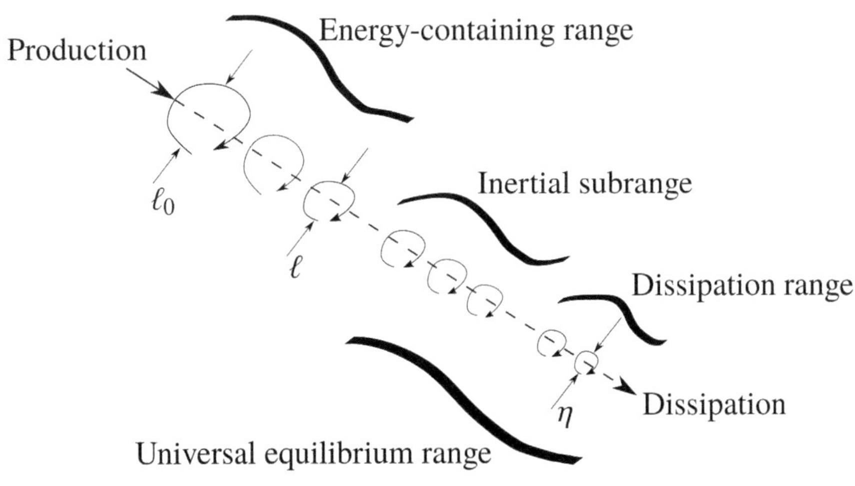

Fig. 6.6 Illustration of the turbulent energy cascade

Starting at the left of Fig. 6.6, we see that kinetic energy is injected into large eddies via turbulence production [28]. These are largest eddies formed in a turbulent flow. They have size ℓ_0, as shown in Fig. 6.6, and this is of the order of the largest scale of the object creating them (for example, the boundary layer height, δ, or cylinder diameter, D). The largest eddies are typically *anisotropic*, meaning that the turbulence intensities in each direction are not the same. The energy cascade describes how these large energy containing eddies breakdown and dissipate their energy.

The energy cascade relies on the fact that large eddies are unstable and will break down into smaller eddies, transferring their energy to these smaller eddies during each breakdown event. The eddies continue breaking down until a limit is reached where the eddies stabilise and dissipate the energy through viscous shear. These small eddies represent the smallest scales of the turbulent flow and are known as the *Kolmogorov* scale, η, after the Russian scientist who first identified them [17].

Kolmogorov shows that as the scale-reduction process continues from the largest eddies, turbulence reaches a point where it becomes locally *isotropic*, meaning that the turbulence intensities are equal in each direction. When this occurs, the eddies enter the so-called *universal equilibrium range* where the statistics of turbulence becomes universal [28]. Energy transfer into the universal equilibrium range is controlled by the largest eddies and the rate they transfer their energy to the equilibrium range. This is easy to understand if we think that energy must be conserved: whatever energy enters the cascade at the largest scales must be dissipated at the smallest scales. The rate of dissipation of turbulent energy is simply known as the dissipation, ϵ, and this process occurs at the right of Fig. 6.6.

Using these concepts, the Kolmogorov scales [17] can be deduced to describe the smallest scales of turbulence, responsible for the dissipation of energy. First, we relate the dissipation to the largest eddies (with velocity and length scales, u_0 and ℓ_0 respectively, see Fig. 6.6) using the scaling result [28],

$$\epsilon \sim \frac{u_o^2}{\tau_0} \sim \frac{u_0^3}{\ell_0} \tag{6.45}$$

which relies on the time scale defined in Eq. (6.44).

The Kolmogorov scales can be deduced using dimensional analysis [28, 36],

$$\eta \equiv \left(\frac{\nu^3}{\epsilon}\right)^{1/4} \tag{6.46}$$

$$u_\eta \equiv (\epsilon\nu)^{1/4} \tag{6.47}$$

$$\tau_\eta \equiv \left(\frac{\nu}{\epsilon}\right)^{1/2} \tag{6.48}$$

where η, u_τ and τ_η are the smallest length, velocity and time scales in the energy cascade.

The universal equilibrium range can be further divided into two further ranges. These are the *inertial subrange* and the *dissipation range*, as shown in Fig. 6.6. In the dissipation range, viscous effects dominate and the major function of this range is to dissipate turbulent energy by viscous action and convert it to heat. At larger scales in the universal equilibrium range, the statistics remain universal but viscous effects are not dominant. Rather, inertial effects control the motion of the eddies and transfer of energy is largely inviscid. For these reasons, this range is called the inertial subrange.

6.5 Statistical Quantities

6.5.1 Mean and Variance

Turbulent flows are random and require statistical tools to describe them. The flow velocity possesses mean and fluctuating quantities. The velocity at a point $\mathbf{x}$ is decomposed as follows,

$$v_i(\mathbf{x}, t) = U_i(\mathbf{x}) + u_i(\mathbf{x}, t) \tag{6.49}$$

where U_i is the mean velocity and u_i is the fluctuating component. The mean or *expected value*, $\mathrm{E}[v_i]$, is defined,

$$U_i = \mathrm{E}[v_i] = \int_{-\infty}^{\infty} v_i\, p(v_i)\; \mathrm{d}v_i \tag{6.50}$$

where $p(v_i)$ is the probability density function of v_i [2]. Similarly, the *variance* is defined as,

$$\mathrm{var}(v_i) = \mathrm{E}[(v_i - U_i)^2] = \int_{-\infty}^{\infty} (v_i - U_i)^2\, p(v_i)\; \mathrm{d}v_i \tag{6.51}$$

Usually, we do not know what the probability density function is, making the evaluation of Eqs. 6.50 and 6.51 impossible. Instead, we often possess a sample of the data over a finite time interval T. We can estimate the mean from this limited sample (which may be a function of time),

$$U_i(t) = \frac{1}{T} \int_{-T/2}^{T/2} v_i(t)\; \mathrm{d}t \tag{6.52}$$

$$\mathrm{var}(v_i(t)) = \overline{u_i^2} = \int_{-T/2}^{T/2} (v_i(t) - U_i(t))^2\; \mathrm{d}t \tag{6.53}$$

where $u_i = v_i - U_i$. The *ensemble average* is used to describe a flow that is repeating over a period of time. For a flow repeated N times,

$$U_{i,N}(t) = \frac{1}{N} \sum_{n=1}^{N} v_i^{(n)}(t) \tag{6.54}$$

where $v_i^{(n)}(t)$ is the nth sample of v_i at time t.

6.5.2 Time Correlations and Spectral Densities

Correlations are important statistical descriptors of random signals and are very useful for analysing turbulent flows. The *autocorrelation* measures the temporal correlation of a signal with itself. For a velocity signal $v(t)$, it is defined,

$$R_{vv}(\tau) = \overline{v(t)v(t+\tau)} = \frac{1}{T} \int_{-T/2}^{T/2} v(t)v(t+\tau)\ \mathrm{d}t \tag{6.55}$$

where τ is the time-delay the correlation is evaluated at. Hence the autocorrelation is simply the average of a signal multiplied by itself at a later time [10].

The *cross-correlation* measures the temporal correlation of two random signals. For two velocity signals $v_1(t)$ and $v_2(t)$, we define it as,

$$R_{v_1 v_2}(\tau) = \overline{v_1(t)v_2(t+\tau)} = \frac{1}{T} \int_{-T/2}^{T/2} v_1(t)v_2(t+\tau)\ \mathrm{d}t \tag{6.56}$$

The *autospectrum* or *power spectral density* is defined as the Fourier transform of the autocorrelation function. For the random velocity signal $v(t)$, the autospectrum is,

$$\phi_{vv}(\omega) = \frac{1}{2\pi} \int_{-T/2}^{T/2} R_{vv}(\tau) e^{i\omega\tau}\ \mathrm{d}\tau \tag{6.57}$$

It can be seen that the autospectrum forms a Fourier transform pair with the autocorrelation function,

$$R_{vv}(\tau) = \int_{-\infty}^{\infty} \phi_{vv}(\omega) e^{-i\omega\tau}\ \mathrm{d}\omega \tag{6.58}$$

The autospectrum $\phi_{vv}(\omega)$ is known as a *two-sided* spectrum because the frequencies are spread equally over both positive and negative ranges, $-\infty \leq \omega \leq \infty$. In practice, we are interested in the positive frequencies only, so we define the *one-sided* spectrum [2],

$$G_{vv}(\omega) = 2\phi_{vv}(\omega) \quad 0 < \omega < \infty \quad \text{otherwise zero} \tag{6.59}$$

The mean square (variance) can be obtained from both the autocorrelation function and the autospectrum,

$$\overline{v^2} = R_{vv}(0) = \int_{-\infty}^{\infty} \phi_{vv}(\omega)\, \mathrm{d}\omega \tag{6.60}$$

which says that the value of the autocorrelation function at zero time delay is the mean square value. This is easy to realise when inspecting the definition of R_{vv} in Eq. (6.58); when $\tau = 0$, the definition of variance is recovered (see Eq. 6.53).

Equation 6.60 also says that the area under the autospectrum is the mean square value. This is because ϕ_{vv} is a power spectral density. In other words, it has the units of {variance of v} per ω. Thus integrating over all ω recovers the variance. Importantly, we can now appreciate that the autospectrum represents the distribution or "power" (v^2 in our example) over different frequencies.

We define the *autocorrelation coefficient* as,

$$\rho(\tau) = \frac{\overline{v(t)v(t+\tau)}}{\overline{v^2}} = \frac{R_{vv}(\tau)}{\overline{v^2}} = \rho(-\tau) \tag{6.61}$$

This is now an even function that has a maximum of unity at $\tau = 0$. We can integrate $\rho(\tau)$ to estimate the *integral time scale* of a turbulent flow,

$$\mathcal{T} = \int_{0}^{\infty} \rho(\tau)\, \mathrm{d}\tau \tag{6.62}$$

The value of $\mathcal{T}$ is a rough measure of the time interval over which $v(t)$ is correlated with itself [35]. Combining the integral time scale with *Taylor's hypothesis*,[1] which states that the turbulence convects at the rate as the mean velocity field. This allows the simple time-space transformation,

$$t \sim \frac{x}{U_c} \tag{6.63}$$

where U_c is the local mean (convection) velocity of the turbulence. Thus, an *integral length scale*, $\mathcal{L}$ can be determined using,

$$\mathcal{L} = \mathcal{T} U_c \tag{6.64}$$

$\mathcal{L}$ is a rough measure of the streamwise lengthscale of the flow.

Similar to the autospectrum, we can define the cross-spectral density as the Fourier transform of the cross-correlation (Eq. 6.56),

[1] Also known as the frozen turbulence assumption.

$$\phi_{v_1v_2}(\omega) = \frac{1}{2\pi}\int_{-T/2}^{T/2} R_{v_1v_2}(\tau)e^{i\omega\tau}\,\mathrm{d}\tau \tag{6.65}$$

and the inverse Fourier transform will return the cross-correlation,

$$R_{v_1v_2} = \int_{-\infty}^{\infty} \phi_{v_1v_2}(\omega)e^{-i\omega\tau}\,\mathrm{d}\omega \tag{6.66}$$

The covariance is,

$$\overline{v_1v_2} = R_{v_1v_2}(\tau = 0) \tag{6.67}$$

To conclude this subsection, we note that the cross-spectral density can be defined using (finite) Fourier transforms [2, 10],

$$\phi_{v_1v_2}(\omega) = \frac{\pi}{T}E[V_1(\omega)V_2^*(\omega)] \tag{6.68}$$

where $V_1(\omega)$ and $V_2(\omega)$ are the Fourier transforms of finite signals $v_1(t)$ and $v_2(t)$ respectively, and $V_2^*(\omega)$ is the complex conjugate of $V_2(\omega)$. Note that the autospectrum can be defined similarly by setting $v_2(t) = v_1(t)$,

$$\phi_{v_1v_1}(\omega) = \frac{\pi}{T}E[V_1(\omega)V_1^*(\omega)] \tag{6.69}$$

6.5.3 Coherence

The *coherence* between two signals $v_1(t)$ and $v_2(t)$ will provide understanding of the connection between them. For example, imagine a turbulent eddy moving through space and passing over two probes spaced a small but finite distance apart. We know that turbulent eddies are unstable and breakdown over time; therefore, we can expect that signal $v_1(t)$ will have a certain level of correlation with $v_2(t)$ and that this level of correlation will diminish as the probes are moved further apart. The coherence is a normalised value that measures the linear connection between these signals. It is defined,

$$\gamma^2_{v_1v_2}(\omega) = \frac{|\phi_{v_1v_2}(\omega)|^2}{\phi_{v_1v_1}(\omega)\phi_{v_2v_2}(\omega)} \tag{6.70}$$

The coherence has values over the range $0 \leq \gamma^2_{v_1v_2} \leq 1$. When $\gamma^2_{v_1v_2} = 0$, then the two signals are uncorrelated and there is no linear relationship between them. When $\gamma^2_{v_1v_2} = 1$, then the signals are perfectly correlated and have linear dependency. When the coherence is greater than zero and less than unity, then one of three situations may exist [2],

1. There is a partial linear connection between the signals. If we interpret the situation as a system with input $v_1(t)$ and output $v_2(t)$, then a coherence between 0 and 1 means the output $v_2(t)$ is caused by input $v_1(t)$ *as well as other inputs.*
2. The system relating the signals is non-linear.
3. There is extraneous noise in the measurements.

For linear systems (such as an acoustic system), the coherence can be interpreted as the fractional portion of the mean-square value at the output ($v_2(t)$) that is contributed by the input ($v_1(t)$) [2].

Returning to our example of two turbulent eddies, the coherence will change from 1 to 0 as the probe separation distance changes from 0 to much greater than the streamwise lengthscale, $\mathcal{L}$ (Eq. 6.64). The astute may realise that turbulence is non-linear making the coherence less of a measure of linearity and more a normalised correlation.

Another pertinent and linear example relates to acoustics. Consider a pressure measurement within the turbulent flow $p_f(t)$ (the input) and a simultaneous acoustic measurement at a point in the far-field $p'(t)$ (the output). The the coherence between these points will measure the fractional portion of the output energy (mean-square acoustic pressure p') due to the input (mean-square turbulent pressure $p_f(t)$). This is a powerful technique that can be used to diagnose acoustic sources in using random data measurements.

Similarly, the phase spectrum is defined,

$$\theta_{v_1 v_2}(\omega) = \arctan\left[\frac{\mathrm{Im}(\phi_{v_1 v_2})}{\mathrm{Re}(\phi_{v_1 v_2})}\right] \tag{6.71}$$

where Re and Im refer to the real and imaginary components of the cross-speatral density. The phase can be related to the time delay. In a linear (or assumed linear) system, we can estimate the time delay as τ_d [2],

$$\tau_d(\omega) = \frac{\theta_{v_1 v_2}(\omega)}{\omega} \tag{6.72}$$

6.5.4 Spatial Correlations

Spatial correlations are used to estimate turbulent length scales and the *wavenumber-frequency* spectra (see Sect. 6.8.4.3). Common applications are for surface pressure loading on elastic plates under turbulent boundary layers, although it is not limited to that particular case.

Spatial correlations are calculated in a similar way as the temporal correlations (see Eq. 6.56). Consider a three-dimensional, time varying field, $a(\mathbf{x}, t)$. For two points separated in the streamwise direction $(x_1, x_1 + r_1)$ at one instant in time and fixed (x_2, x_3), we can define the following streamwise spatial correlation,

$$R(r_1) = E[a(x_1, x_1 + r_1)] \approx \frac{1}{L} \int_{-L/2}^{L/2} a(x_1, x_1 + r_1)\, dx_1 \tag{6.73}$$

which is an integral along a line. Here, L is the streamwise extent over which the field is known. As $L \to \infty$, the integral approaches the true expected value.

This can be converted to a wavenumber spectrum using the spatial Fourier transform,

$$\phi(k_1) = \frac{1}{2\pi} \int_{-\infty}^{\infty} R(r_1) e^{-ik_1 r_1}\, dr_1 \tag{6.74}$$

This concept is used to define spectral models of homogeneous turbulence (Sect. 6.7) and the wavnumber-frequency wall-pressure spectrum (Sect. 6.8.4.3).

6.6 Reynolds Averaging

The process of decomposing turbulent quantities such as velocity into its mean and fluctuating quantities is also known as *Reynolds averaging*, after Osbourne Reynolds, the famous scientist who first described turbulence and transition. Placing the decomposed flow variables into the continuity and Navier-Stokes equations and performing temporal averaging gives the *Reynolds averaged* Navier-Stokes (RANS) and continuity equations. These provide great insight into turbulent flow processes that serve as a foundation to understanding flow noise in later chapters.

First, we decompose the flow variables into mean (expected) and fluctuating values,

$$v_i(\mathbf{x}, t) = U_i(\mathbf{x}) + u_i(\mathbf{x}, t) \tag{6.49}$$

$$p(\mathbf{x}, t) = \overline{p}(\mathbf{x}) + p'(\mathbf{x}, t) \tag{6.75}$$

$$\rho(\mathbf{x}, t) = \overline{\rho}(\mathbf{x}) + \rho'(\mathbf{x}, t) \tag{6.76}$$

where p is pressure, $\overline{p}$ is mean pressure, p' is the fluctuating pressure, $\overline{\rho}$ is mean density, ρ is density and ρ' is the fluctuating density.

We also define the *compressive stress tensor* [1, 10], which combines pressure and viscous stress,

$$p_{ij} = p\delta_{ij} - \sigma_{ij} \tag{6.77}$$

where σ_{ij} is the viscous stress tensor,

$$\sigma_{ij} = \mu \left(\frac{\partial v_i}{\partial x_j} + \frac{\partial v_j}{\partial x_i} - \frac{2}{3} \frac{\partial v_k}{\partial x_k} \delta_{i,j} \right) \tag{6.78}$$

and μ is the dynamic viscosity. The compressive stress tensor can also be decomposed into mean and fluctuating components,

$$p_{ij} = \overline{p_{ij}} + p'_{ij} \tag{6.79}$$

We now write the compressible continuity and Navier-Stokes equations,

$$\frac{\partial \rho}{\partial t} + \frac{\partial(\rho v_i)}{\partial x_i} = 0 \tag{6.80}$$

$$\frac{\partial(\rho v_i)}{\partial t} + \frac{\partial(\rho v_i v_j + p_{ij})}{\partial x_j} = 0 \tag{6.81}$$

Inserting the decomposed flow variables (Eqs. 6.49, 6.75, 6.76 and 6.79) into the above and performing time-averaging (Eq. 6.52) yields the full, compressible Reynolds Averaged continuity and Navier-Stokes equations,

$$\frac{\partial \overline{\rho}}{\partial t} + \frac{\overline{\rho} U_i + \overline{\rho' u_i}}{\partial x_i} = 0 \tag{6.82}$$

$$\frac{\partial}{\partial t}\left[\overline{\rho} U_i + \overline{\rho' u_i}\right] + \frac{\partial}{\partial x_i}\left[\overline{\rho} U_i U_j + \overline{\rho}\,\overline{u_i u_j} + \overline{\rho' u_i u_j} + \overline{\rho' U_i u_j} + \overline{\rho' U_j u_j} + \overline{p_{ij}}\right] = 0 \tag{6.83}$$

To arrive at these equations, we have used the fact that the mean of the fluctuating quantities are zero. For the example of velocity,

$$\overline{u_i} = \frac{1}{T}\int_t^{t+T} u_i(\tau)\, \mathrm{d}\tau = 0 \tag{6.84}$$

However, the mean of the product of fluctuating quantities need not be zero. For example, the average of two orthogonal fluctuating velocities is,

$$\overline{u_i u_j} = \frac{1}{T}\int_t^{t+T} u_i(\tau) u_j(\tau)\, \mathrm{d}\tau \neq 0 \tag{6.85}$$

and as shown, this need not be equal to zero. These averaged quantities measure the correlation between the variables [36]; if $\overline{u_i u_j} = 0$, u_i and u_j are said to be uncorrelated. If $\overline{u_i u_j} \neq 0$, they are said to be correlated.

We can now immediately see the effect of turbulence through the compressible continuity and Reynolds averaged Navier-Stokes equations, Eqs. (6.82) and (6.83). In the continuity equation, Eq. (6.82), the correlation $\overline{\rho' u_i}$ adds an additional mass flux term, which can be interpreted as the action of turbulent flow influencing mass transfer.

In the Reynolds averaged Navier-Stokes equations, Eq. (6.83), we can see the influence on mass transfer in the first term. In the second term, we see a number of additional correlations. These act in the same manner as the compressive stress tensor and are therefore known as turbulent *stresses*. If we take the case of a low Mach number flow,

$$\rho' \ll \overline{\rho} \tag{6.86}$$

the correlations involving ρ' are insignificant compared with those involving mean density in areas of turbulent flow. This means that at low Mach number and high Reynolds number (high enough for turbulent flow), acoustic density fluctuations are very small and do not couple with the turbulent motion. This decoupling of acoustics and turbulence allows the use of acoustic analogy methods (see Sect. 7.2.1) to calculate acoustic fields using incompressible flow field data. In other words, an incompressible turbulent flow simulation can be used to determine the Reynolds stresses. These Reynolds stresses can then be used to calculate the acoustic field. It is one-way coupled, making it an efficient process. Note that for low Reynolds number laminar and transitional flows, this is not the case and a coupled prediction methodology will be needed. This is because in some situations, the acoustic density, velocity and pressure fields are of the same order as the flow field and strong coupling may occur, causing what is sometime called acoustic feedback. This can occur during low Reynolds number flow over cavities and airfoils.

At low Mach number and high Reynolds number, ρ' is insignificant compared with the mean density and the most important stress terms are the *Reynolds stresses*, $\overline{\rho}\,\overline{u_i u_j}$, or dropping the density term, simply, $\overline{u_i u_j}$. The Reynolds stress can be considered a time-averaged rate of momentum transfer due to turbulence. They show how momentum transfer, created by the three-dimensional fluctuating velocity field, effectively adds to the viscous stresses in the fluid.

The Reynolds stress components $\overline{u_i u_j}$ form a symmetric second-order tensor. The diagonal terms, $\overline{u}_1^2 = \overline{u_1 u_1}$, $\overline{u}_2^2$ and $\overline{u}_3^2$ are known as the *normal stresses*. The symmetric, off-diagonal terms, $\overline{u_i u_j} = \overline{u_j u_i}$ are called the *shear stresses*. For reference, the full Reynolds stress tensor is written,

$$\overline{u_i u_j} = \begin{bmatrix} \overline{u}_1^2 & \overline{u_1 u_2} & \overline{u_1 u_3} \\ \overline{u_2 u_1} & \overline{u}_2^2 & \overline{u_2 u_3} \\ \overline{u_3 u_1} & \overline{u_3 u_2} & \overline{u}_3^2 \end{bmatrix} = \begin{bmatrix} \overline{u}_1^2 & \overline{u_1 u_2} & \overline{u_1 u_3} \\ \overline{u_1 u_2} & \overline{u}_2^2 & \overline{u_2 u_3} \\ \overline{u_1 u_3} & \overline{u_2 u_3} & \overline{u}_3^2 \end{bmatrix} \tag{6.87}$$

Predicting what these Reynolds stress correlations should be is a central focus of turbulence modelling [36] and remains an unresolved question in fluid mechanics [28]. Modern simulation tools, such as direct numerical simulation (DNS) and large eddy simulation (LES) attempt to model all or nearly all of the turbulent scales directly, thus provide estimates of the Reynolds stresses; however, experimental validation of these methods, particularly at high Reynolds number, remains a challenge. Experimental measurements of turbulence are therefore necessary if we are to further our understanding of turbulence. Additionally, the Reynolds stresses also form the principal components of turbulent flow noise sources (see Sect. 7.2.1).

When the Mach number is high enough for compressibility to become important (usually when the freestream Mach number is $M_\infty > 0.3$), then the density fluctuations (ρ') are significant. This means that the mass flux terms in Eqs. (6.82) and (6.83) cannot be ignored. Further, the correlations involving ρ' in the second term of Eq. (6.83) add additional turbulent stress to the momentum equation. Hence, strong compressibility alters how turbulence influences the flow field and increases the complexity of the processes involved. In fact, a *triple correlation*, $\overline{\rho' u_i u_j}$, plays an important role. Measuring and predicting such a correlation is rather difficult. For turbulence modelling, the density-weighted or Favre-averaged equations [36] are used to reduce the complexity of numerical modelling.

6.6.1 Turbulent Kinetic Energy and Anisotropy

The *turbulent kinetic energy*, k, is defined as half the sum of the diagonal terms (the trace) of the Reynolds stress tensor [28],

$$k = \frac{1}{2}\overline{u_i u_i} \tag{6.88}$$

The turbulence intensities (TI_i) for each velocity component are defined as [36],

$$TI_i = \frac{\sqrt{\overline{u_i u_i}}}{U_\infty} \tag{6.89}$$

where U_∞ is the freestream velocity. This is sometimes replaced by the external velocity U_e for boundary layers. In the special case of isotropic turbulence, where,

$$\overline{u_1'^2} \approx \overline{u_2'^2} \approx \overline{u_3'^2} \tag{6.90}$$

a general turbulence intensity can be defined using the turbulent kinetic energy,

$$TI = \sqrt{\frac{2}{3}\frac{k^2}{U_\infty^2}} \tag{6.91}$$

Equation (6.91) is particularly useful for relating the turbulent kinetic energy obtained from turbulence models to a normal stress.

When turbulent flow is not isotropic and contains significant shear stress values in the off-diagonal terms of the Reynolds stress tensor, we say the turbulence is *anisotropic*. The normal stresses can be assembled as an isotropic stress tensor,

$$\overline{u_i u_i} = \tfrac{2}{3}k\delta_{ij} \tag{6.92}$$

Then, the deviatoric stress component part of the Reynolds stress tensor is [28],

$$a_{ij} = \overline{u_i u_j} - \tfrac{2}{3}k\delta_{ij} \tag{6.93}$$

We can now see that is this anisotropic component that is responsible for momentum transport in turbulence. For clarity, we assume low Mach number flow ($\rho' \ll \overline{\rho}$) and isolate the second term of Eq. (6.83),

$$\begin{aligned} &\overline{\rho}U_iU_j + \overline{\rho}\,\overline{u_iu_j} + \overline{p_{ij}} \\ &= \overline{\rho}U_iU_j + \overline{\rho}a_{ij} + (\overline{p_{ij}} + \tfrac{2}{3}\overline{\rho}k\delta_{ij}) \end{aligned} \tag{6.94}$$

Equation (6.94) tells us that momentum transport has a mean convection part ($\overline{\rho}U_iU_j$) and a turbulent part. In the turbulent part, the isotropic stresses act just like pressure and can be absorbed into the pressure term, $(\overline{p_{ij}} + \frac{2}{3}\overline{\rho}k\delta_{ij})$. The remaining anisotropic stresses (a_{ij}) are therefore responsible for enhanced momentum transport in turbulent flow.

6.7 Spectral Models for Homogeneous Isotropic Turbulence

We now turn our attention to the wavenumber spectrum, $\phi_{ij}(\mathbf{k})$ where $\mathbf{k} = (k_1, k_2, k_3)$ is the wavenumber vector. We will first define wavenumbers, the wavenumber-frequency spectrum and the integral length scale before listing common analytical relationships for isotropic turbulence spectra.

Simply, the wavenumber represents the wavelength (λ_i) in each direction,

$$k_i = \frac{2\pi}{\lambda_i} \tag{6.95}$$

If the waves convect with constant velocity U_{c_i} in the ith direction for frequency ω, then we can also write,

$$k_i = \frac{\omega}{U_{c_i}} \tag{6.96}$$

To define the wavenumber-frequency spectrum, we first need to define the time correlation tensor for velocity at two points separated in space by $\mathbf{r} = x_i - y_i$ is,

$$R_{ij}(\mathbf{r}, \tau) = \frac{1}{T}\int_{-T/2}^{T/2} u_i(x_i, t)u_j(y_i, t+\tau)\,\mathrm{d}t \tag{6.97}$$

The wavenumber-frequency spectrum of u_iu_j is therefore,

$$\phi_{ij}(\mathbf{k},\omega) = \frac{1}{(2\pi)^4}\int_{-T/2}^{T/2}\int_{-\infty}^{\infty}\int_{-\infty}^{\infty}\int_{-\infty}^{\infty} R_{ij}(\mathbf{r},\tau)\exp\left[i(\omega\tau - \mathbf{k}\cdot\mathbf{r})\right]\,\mathrm{d}^3\mathbf{r}\mathrm{d}\tau \tag{6.98}$$

which reduces to,

$$\phi_{ij}(\mathbf{k},\omega) = \frac{1}{(2\pi)^3}\int_{-\infty}^{\infty}\int_{-\infty}^{\infty}\int_{-\infty}^{\infty} \phi_{ij}(\mathbf{r},\omega)\exp\left[-i\mathbf{k}\cdot\mathbf{r}\right]\,\mathrm{d}^3\mathbf{r} \tag{6.99}$$

We now use Taylor's frozen flow hypothesis (Sect. 6.5.2), that assumes the turbulence convects with velocity $\mathbf{U_c}$, to separate the wavenumber and frequency components,

$$\phi_{ij}(\mathbf{k},\omega) = \phi_{ij}(\mathbf{k})\delta(\omega - \mathbf{U_c}\cdot\mathbf{k}) \tag{6.100}$$

We therefore have defined the wavenumber-frequency spectrum and can now move on to presenting analytical models of the turbulence velocity spectrum.

In isotropic turbulence, the wavenumber spectrum can be written [3, 28],

$$\phi_{ij}(\mathbf{k}) = \frac{E(k)}{4\pi k^4}\left[k^2\delta_{ij} - k_i k_j\right] \tag{6.101}$$

where

$$k^2 = |\mathbf{k}|^2 = k_1^2 + k_2^2 + k_3^2 \tag{6.102}$$

Note that the wavenumber uses the same symbol as turbulent kinetic energy used previously, which is confusing.

The one-dimensional energy spectrum $E(k)$ [28] is used to calculate the wavenumber spectra and relate this to key properties of the flow. For isotropic turbulence, the autospectrum of streamwise velocity fluctuations can be determined using the following identity [3, 28],

$$E_1(k_1) = 2\int_{-\infty}^{\infty}\int_{-\infty}^{\infty} \phi_{11}(\mathbf{k})\,\mathrm{d}k_2\mathrm{d}k_3 = 2\phi_{11}(k_1) \tag{6.103}$$

We now can define the length Λ_f as the integral length scale of the flow (similar to the estimate provided in Sect. 6.5.2,

$$2\Lambda_f = \frac{1}{\overline{u_1^2}}\int_{-\infty}^{\infty} R_{11}(r_1)\,\mathrm{d}r_1 \tag{6.104}$$

where R_{11} is defined in Eq. (6.97). As the flow is isotropic, we can write,

$$\overline{u_1^2} = \int_0^{\infty} E_1(k_1)\,\mathrm{d}k_1 = \overline{u_2^2} = \overline{u_3^2} = \overline{u^2} \tag{6.105}$$

We now can use these expressions to write analytical functions for the energy and streamwise velocity autospectra.

6.7.1 von Karman Spectrum

The von Karman spectrum is commonly used and is,

$$E(k) = \frac{55}{9} \frac{\Gamma(5/6)}{\sqrt{\pi}\Gamma(1/3)} \frac{\overline{u^2}}{k_e} \frac{(k/k_e)^4}{[1 + (k/k_e)^2]^{5/6}} \tag{6.106}$$

where $\Gamma(x)$ is the gamma function evaluated at x and k_e is an inverse scale that characterises the energy containing eddies,

$$k_e = \frac{\sqrt{\pi}}{\Lambda_f} \frac{\Gamma(5/6)}{\Gamma(1/3)} \approx \frac{3}{4\Lambda_f} \tag{6.107}$$

Using Eq. (6.103), we can find the velocity autospectrum,

$$E_1(k_1) = \frac{2}{\sqrt{\pi}} \frac{\Gamma(5/6)}{\Gamma(1/3)} \frac{\overline{u^2}}{k_e} \frac{1}{[1 + (k/k_e)^2]^{5/6}} \tag{6.108}$$

6.7.2 Liepmann Spectrum

The Liepmann spectrum is [19–21],

$$E(k) = \frac{8}{\pi} \overline{u^2} \Lambda_f \frac{(k\Lambda_f)^4}{[1 + (k\Lambda_f)^2]^3} \tag{6.109}$$

6.8 Turbulent Boundary Layers

6.8.1 Power Law Profiles

An approximation of turbulent boundary layer velocity profiles can be given as a power law [32, 36],

$$\frac{u}{U_e} = \left(\frac{y}{\delta}\right)^{1/n} \tag{6.110}$$

where U_e is the velocity external to the boundary layer and is often taken as the freestream velocity U_∞ for a flat plate, zero-pressure-gradient boundary layer; y is the distance normal to the surface where the boundary layer height δ is measured. The exponent is typically taken to be $n = 7$ for a high Reynolds number boundary layer ($Re_x \gtrsim 10^7$, but often it works for tripped boundary layers at lower Reynolds numbers).

6.8.2 *Law of the Wall*

Power law velocity profiles can only be taken as approximate. More accurate is the *law of wall*, a foundation of turbulent boundary layer theory. The law of the wall rests on observations that show that the streamwise mean velocity in a turbulent boundary layer near the wall (but not at the wall) vary logarithmically with height. Not that unreasonable an assumption when power law models provide a reasonable fit to boundary layer velocity data. The law of the wall improves on this by using dimensional analysis. For a smooth wall with turbulent boundary layer with fluid density ρ, we can suppose that the wall stress, τ_w, can be expressed in terms of a *friction velocity*, u_τ,

$$\tau_w = \rho u_\tau^2 \tag{6.111}$$

Rearranging, we have the formal definition of friction velocity,

$$u_\tau \equiv \sqrt{\frac{\tau_w}{\rho}} \tag{6.112}$$

Dimensional analysis yields [13],

$$\frac{\partial u}{\partial y} = \frac{u_\tau}{y}\mathcal{F}\left(\frac{u_\tau y}{\nu}\right) \tag{6.113}$$

where $\mathcal{F}\left(\frac{u_\tau y}{\nu}\right)$ is a function that approaches a constant as $\frac{u_\tau y}{\nu} \to \infty$, which is based on experimental observations and the reasoning that the mean flow will not interact with the large eddies in the boundary layer at high Reynolds number [13]. We define this constant as the inverse of Kàrman's constant, κ, so that we can now write Eq. (6.113) as,

$$\frac{\partial u}{\partial y} = \frac{u_\tau}{\kappa y} \tag{6.114}$$

Defining the non-dimensional numbers,

$$u^+ = \frac{u}{u_\tau} \tag{6.115}$$

$$y^+ = \frac{u_\tau y}{\nu} \tag{6.116}$$

which allows us to non-dimensionalise Eq. (6.114),

$$\frac{\partial u^+}{\partial y^+} = \frac{1}{\kappa y^+} \tag{6.117}$$

Integrating Eq. (6.117) over y^+ yields the logarithmic *law of the wall*,

$$u^+ = \frac{1}{\kappa} \ln y^+ + C \tag{6.118}$$

where the constants have been empirically determined as $\kappa = 0.41$ and $C \approx 5$.

Figure 6.7 shows the experimental mean streamwise velocity data of Hutchins et al. [14] expressed in terms of y^+ and u^+. These data were obtained at a friction Reynolds number of $Re_\tau \sim 14,000$, where $Re_\tau = \frac{\delta u_\tau}{\nu}$. Above $y^+ = 30$, the experimental data follows a logarithmic law, confirmed by the excellent comparison with the law of the wall (Eq. 6.118). This region is called the log layer (as shown in

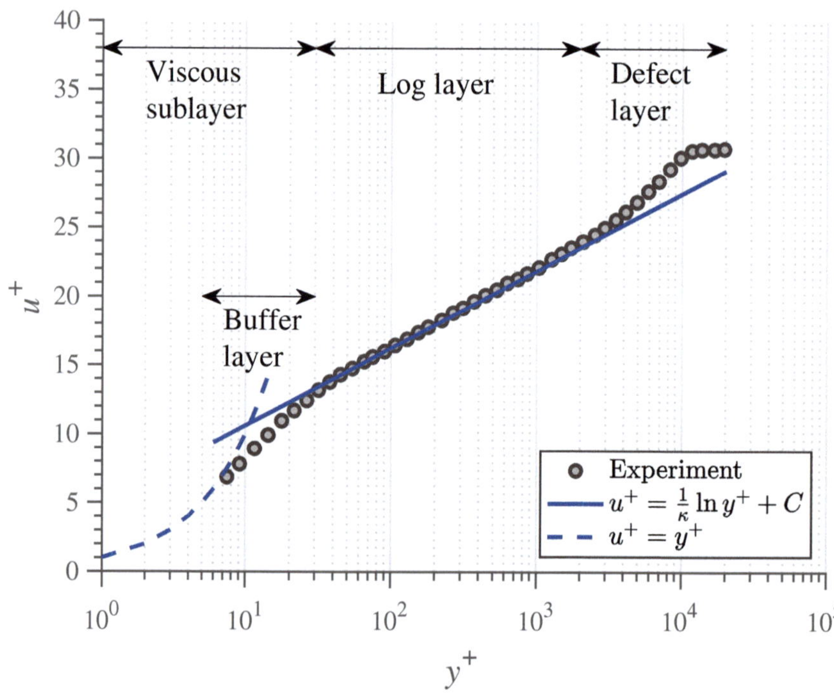

Fig. 6.7 Velocity profile in a zero-press-gradient turbulent boundary layer compared with the law of the wall. Experimental data from Hutchins et al. [14] for $Re_\tau \sim 14{,}000$; $\kappa = 0.41$ and $C = 5$

Fig. 6.7) and extends from $y^+ > 30$ to approximately $\frac{y}{\delta} = 0.2$. The region close to the wall for $y+ \leq 30$ is called the viscous sublayer. It is composed of two regions. For $y^+ \leq 5$, the linear sublayer holds. This can be thought of a very small laminar zone and can be described by,

$$u^+ = y^+ \tag{6.119}$$

and this equation is the dashed line in Fig. 6.7. In the region $5 < y^+ \leq 30$ we have the buffer layer, which is the transition zone between the linear sublayer and the log layer.

For $\frac{y}{\delta} \gtrsim 0.2$ we have the defect layer (as shown in Fig. 6.7), sometimes known as the outer layer, and this extends to the edge of the boundary layer. As shown in Fig. 6.7, the velocity profile departs significantly from the logarithmic relationship of Eq. (6.118). The law of the wall can be modified in to the composite *law of the wake*, first determined by Coles [6],

$$u^+ = \frac{1}{\kappa} \ln y^+ + C + \frac{2\Pi}{\kappa} \sin^2\left(\frac{\pi}{2}\frac{y}{\delta}\right) \tag{6.120}$$

Here, the right-hand-term of Eq. (6.120) adds a perturbation to the defect layer that is well-correlated with experiments. The new parameter, Π, is Coles' wake-strength parameter and is found to be $\Pi \sim 0.51 - 0.6$ for zero-pressure-gradient boundary layers but is sensitive to pressure gradient. The equilibrium parameter [36] is a non-dimensional pressure gradient term,

$$\beta_T \equiv \frac{\delta^*}{\tau_w}\frac{\mathrm{d}\overline{p}}{\mathrm{d}x} \tag{6.121}$$

Using data from Skåre and Krogstad [33] and Coles and Hirst [7], Wilcox et al. [36] present the following empirical model that relates Coles' wake parameter with the equilibrium parameter,

$$\Pi = 0.6 + 0.51\beta_T - 0.01\beta_T^2 \tag{6.122}$$

6.8.3 *Flat Plate Empirical Relationships*

Consider a turbulent boundary layer formed on a flat plate at zero pressure gradient. The boundary layer thickness can be calculated using the empirical relationship,

$$\delta = \frac{0.385}{Re_x^{0.2}} \tag{6.123}$$

where Re_x is the Reynolds number based on the distance from the leading edge of the flat plate.

An empirical model for displacement thickness is,

$$\delta^* = \frac{0.048}{Re_x^{0.2}} \tag{6.124}$$

Similarly, the momentum thickness is,

$$\theta = \frac{0.037}{Re_x^{0.2}} \tag{6.125}$$

The skin-friction coefficient is,

$$c_f(x) = \frac{0.0594}{Re_x^{0.2}} \tag{6.126}$$

The Schulz-Grunow empirical formula relates skin friction to Reynolds number,

$$c_f(x) = 0.37(\log_{10} Re_x)^{-2.584} \tag{6.127}$$

6.8.4 Turbulent Wall-Pressure Fluctuations

When a turbulent boundary layer passes over a solid surface, such as an airfoil, it induces fluctuations in pressure at the surface, which we call wall-pressure fluctuations. These are responsible for flow-induced noise and vibration.

We describe the wall-pressure as having a mean ($\overline{p}$) and fluctuating (p') components as described in Eq. (6.75). These pressure fluctuations are induced by the turbulence above the surface. Assuming incompressible flow, we can rearrange the Navier Stokes equations by taking the divergence and using Reynolds averaging to give [5, 30],

$$\frac{1}{\rho_0}\frac{\partial^2 p'}{\partial x_i^2} = 2\frac{\partial U_i}{\partial x_j}\frac{\partial u_j}{\partial x_i} - \frac{\partial^2\left(u_i u_j - \overline{u_i u_j}\right)}{\partial x_i \partial x_j} \tag{6.128}$$

An integral solution for the fluctuating wall pressure can be found using a flat-plate Green's function [27, 34],

$$p'(\mathbf{x}, t) = \frac{\rho_0}{2\pi}\int_V \left[2\frac{\partial U_1}{\partial y_2}\frac{\partial u_2}{\partial y_1} + \frac{\partial^2(u_i u_j - \overline{u_i u_j})}{\partial y_i \partial y_j}\right]\frac{\mathrm{d}V(\mathbf{y})}{|\mathbf{x}-\mathbf{y}|} \tag{6.129}$$

where $\mathbf{y}$ is the position of the turbulent source in the boundary layer and $V(\mathbf{y})$ is the volume over which the integration is performed, ideally but not practically, the volume is infinite. We can immediately see from Eq. (6.129) the wall pressure at a point $\mathbf{x}$ is created by turbulence everywhere in volume $V(\mathbf{y})$. Further, the wall pressure has two main components, a turbulence-mean-shear (TMS) interaction term,

$$2\frac{\partial U_1}{\partial y_2}\frac{\partial u_2}{\partial y_1}$$

and a turbulence-turbulence (TT) interaction term,

$$\frac{\partial^2(u_i u_j - \overline{u_i u_j})}{\partial y_i \partial y_j}$$

This representation provides a useful framework for understanding the production of wall pressure fluctuations. It shows that the interaction of mean velocity shear with turbulence (TMS) induces wall pressure fluctuations. As this can occur over a short time scales, it is often thought of as a 'fast' acting component. The TT interaction is due to the interaction of turbulence with itself and often occurs over a longer timescale than the TMS term, thus is sometimes regarded as a 'slow' acting component. Regardless, it can be seen that the fluid mechanics in the volume about $\mathbf{x}$ is important, including both the mean and turbulent fields.

6.8.4.1 RMS Pressure Levels

Miller et al. [25] provide a comprehensive review of turbulent boundary layer RMS wall-pressure empirical models. They find, after comparison with their own experimental data, that most RMS models estimate RMS levels reasonably below a Mach number of 0.3; however, there is disagreement at higher Mach numbers. In the absence of further guidance, two RMS wall-pressure level models will be listed here. The first is the shear stress-Reynolds number dependent model of Farabee and Casarella [9],

$$\overline{p'^2} = 6.5\tau_w^2 \qquad \text{for } Re_\tau \leq 333 \tag{6.130}$$

$$\overline{p'^2} = 6.5 + 1.86\ln(Re_\tau/333) \qquad \text{for } Re_\tau > 333 \tag{6.131}$$

where τ_w is the shear stress at the measuring position at the wall (under the boundary layer) and the Reynolds number Re_τ is defined,

$$Re_\tau = \frac{\delta u_\tau}{\nu} \tag{6.132}$$

where δ is the boundary layer thickness, u_τ is the friction wall velocity (defined $u_\tau = \sqrt{(\tau_w/\rho)}$ and ν is the kinematic viscosity of the fluid.

The second empirical model is based on dynamic pressure and was developed by Lueptow [23]. It is simply,

$$\overline{p'^2} \approx (0.012q)^2 \tag{6.133}$$

where q is the dynamic pressure based on the fluid velocity external to the boundary layer. As noted by Miller et al. [25], there is no agreement on what is the best RMS wall pressure model at the present time. However, each of these models provides a reasonable estimate and a choice can be made based on the convenience of using either the Reynolds number or dynamic pressure.

6.8.4.2 Empirical Wall-Pressure Single-Point Spectrum Model

There are many flate-plate, zero-pressure gradient wall-pressure spectrum empirical models available in the literature. Recent papers [18, 25] summarise the performance of flat plate, zero pressure gradient wall-pressure single-point spectral models. On reflection, it has been shown that the Goody model [11] is the most appropriate to use; however, there are still inaccuracies and more work is required.

The Goody model is [11],

$$\frac{\phi(\omega)U_e}{\tau_w^2\delta} = \frac{3.0(\omega\delta/U_e)^2}{[(\omega\delta/U_e)^{0.75}+0.5]^{3.7}+[(1.1R_T^{-0.57})(\omega\delta/U_e)]^7} \tag{6.134}$$

where U_e is the velocity at the edge of the boundary layer and R_T is the ratio of the outer-layer-to-inner-layer timescale,

$$R_T = \frac{u_\tau\delta}{\nu}\sqrt{\frac{c_f}{2}} \tag{6.135}$$

6.8.4.3 Wavenumber-Frequency Wall-Pressure Spectrum

The temporal and spatial statistics of turbulent boundary layer wall-pressure is a source of flow induced noise. It can create direct noise, which is weak at low Mach numbers, or can be transformed to more significant radiated noise by edge scattering or wall vibration.

Consider a turbulent boundary layer on a flat plate formed under a zero pressure gradient condition. We set the streamwise direction as x, the wall normal direction as y and the spanwise direction as z. We also use the tensor notation so that $(x, y, z) = x_i$ and the wavenumber vector is represented by $\mathbf{k} = (k_1, k_2, k_3) = k_i$.

The wavenumber-frequency spectrum is a useful statistical tool that is used to describe the boundary layer pressure excitation at the wall. It conveniently shows the wavelength (for each orthogonal direction) of the turbulent pressure field at each frequency, which is important for noise and vibration analysis. The wavenumber-frequency spectrum is defined as,

$$\phi(k_1, k_3, \omega) = \frac{1}{(2\pi)^3} \int_{-T/2}^{T/2} \int_{-\infty}^{\infty} \int_{-\infty}^{\infty} R_{pp}(r_1, r_3, \tau) e^{-ik_1 r_1} e^{-ik_3 r 3} e^{i\omega\tau} \, \mathrm{d}r_1 \mathrm{d}r_3 \mathrm{d}\tau \tag{6.136}$$

where $R_{pp}(r_1, r_3, \tau)$ is the two-dimensional space-time correlation,

$$R_{pp}(r_1, r_3, \tau) = \overline{p'(x_1, x_3, t) p'(x_1 + r_1, x_3 + r_3, t + \tau)} \tag{6.137}$$

where r_1 and r_3 are the separation distances in the x_1 and x_3 directions respectively. The time delay is τ. Physically, the flow can be thought of as a random fluctuating field that follows the flow over the plate [25]. To mimic the observations of turbulence, the amplitude of the correlation function decays in an exponential-like manner in both the streamwise and spanwise directions.

Models of the wavenumber-frequency spectrum are empirical because of the intractable nature of turbulence. Two types of model are commonly used, separable and non-separable. A widely used separable model is the Corcos [8], which can be written in spatial (x_1, x_3) form,

$$\phi(r_1, r_3, \omega) = \phi(\omega) \exp\left(ik_\omega x_1 - \alpha_1 |k_\omega r_1| - \alpha_3 |k_\omega r_3|\right) \tag{6.138}$$

Here, $\phi(\omega)$ is a single-point wall-pressure model, such as the Goody model described in Sect. 6.8.4.2. The model is described as separable because the streamwise and spanwise decay rates can be separated from the equation as can the single-point wall-pressure spectrum. The constants α_1 and α_3 are the empirical decay rates in the streamwise and spanwise directions. Taking the Fourier transform in space yields the complete Corcos wavenumber-frequency spectrum,

$$\phi(k_1, k_3, \omega) = \frac{\phi(\omega) U_c^2}{\pi^2 \omega^2} \frac{\alpha_1 \alpha_3}{[\alpha_1^2 + (U_c k_1/\omega - 1)^2][\alpha_3^2 + (U_c k_3/\omega - 1)^2]} \tag{6.139}$$

The values for the empirical decay rates are: $\alpha_1 = 0.1$ and $\alpha_3 = 0.77$.

There are many non-separable wavenumber-frequency wall-pressure spectrum models available as well [12, 25].

6.9 Summary

This chapter provides an overview of laminar and turbulent flow, with a view towards understanding flow-induced noise. Laminar flow cases are used to introduce the chapter. Laminar boundary layer theory is used to present the Blasius equation and subsequent formula to predict thickness and skin friction. A similar approach is used for other canonical flows such as the plane mixing layer, free jet and wake flows.

Transition is defined and analysed through the use of the Orr-Sommerfeld equation. Then turbulent flow is treated in some detail. A physical description of the energy cascade is given first. This is followed by definitions of statistical quantities such as mean, variance, correlations and spectra. Coherence is also defined which is also very useful for acoustic analysis. Reynolds averaging is explained and used as a basis for understanding turbulent flow processes. Spectral models of homogeneous, isotropic turbulence are provided.

Turbulent boundary layers are presented. The law of the wall and law of the wake are described. Empirical relationships for flat plate boundary layer thickness and skin friction coefficient are included. The chapter ends with descriptions and models for turbulent wall pressure.

References

1. Batchelor G (2000) An introduction to fluid dynamics. Cambridge University Press, Cambridge
2. Bendat JS, Piersol AG (2011) Random data: analysis and measurement procedures, vol 729. Wiley
3. Blake WK (2017) Mechanics of flow-induced sound and vibration, vol 1, General concepts and elementary sources, 2nd edn. Academic press, London
4. Browand FK (1966) An experimental investigation of the instability of an incompressible, separated shear layer. J Fluid Mech 26(2):281–307
5. Bull M (1996) Wall-pressure fluctuations beneath turbulent boundary layers: some reflections on forty years of research. J Sound Vib 190(3):299–315
6. Coles D (1956) The law of the wake in the turbulent boundary layer. J Fluid Mech 1(2):191–226. https://doi.org/10.1017/S0022112056000135
7. Coles D, Hirst E (1969) Computation of turbulent boundary layers-1968 afosr-ifp-stanford conference, vol II
8. Corcos G (1964) The structure of the turbulent pressure field in boundary-layer flows. J Fluid Mech 18(3):353–378
9. Farabee TM, Casarella MJ (1991) Spectral features of wall pressure fluctuations beneath turbulent boundary layers. Phys Fluids A 3(10):2410–2420. https://doi.org/10.1063/1.858179, http://aip.scitation.org/doi/10.1063/1.858179
10. Glegg S, Devenport W (2017) Aeroacoustics of low Mach number flows: fundamentals, analysis, and measurement. Academic Press
11. Goody M (2004) Empirical spectral model of surface pressure fluctuations. AIAA J 42(9):1788–1794. https://doi.org/10.2514/1.9433
12. Graham WR (1997) A comparison of models for the wavenumber-frequency spectrum of turbulent boundary layer pressures. J Sound Vib 206(4):541–565. https://doi.org/10.1006/jsvi.1997.1114
13. Huang PG, Bradshaw P (1995) Law of the wall for turbulent flows in pressure gradients. AIAA J 33(4):624–632. https://doi.org/10.2514/3.12624
14. Hutchins N, Nickels TB, Marusic I, Chong MS (????) Hot-wire spatial resolution issues in wall-bounded turbulence. J Fluid Mech 103–136. https://doi.org/10.1017/S0022112009007721
15. Jacobs R, Durbin P (2001) Simulations of bypass transition. J Fluid Mech 428:185–212
16. Kingan MJ, Pearse JR (2009) Laminar boundary layer instability noise produced by an aerofoil. J Sound Vib 322(4–5):808–828
17. Kolmogorov AN (1941) The local structure of turbulence in incompressible viscous fluid for very large Reynolds numbers. Doklady Akademiia Nauk SSSR 30:301–305
18. Lee S, Shum JG (2019) Prediction of airfoil trailing-edge noise using empirical wall-pressure spectrum models. AIAA J 57(3):888–897. https://doi.org/10.2514/1.J057787, https://arc.aiaa.org/doi/10.2514/1.J057787

19. Liepmann H, Laufer J, Liepmann K (1951) On the spectrum of isotropic turbulence
20. Liepmann HW (1952) On the application of statistical concepts to the buffeting problem. J Aeronaut Sci 19(12):793–800
21. Liepmann HW (1955) Extension of the statistical approach to buffeting and gust response of wings of finite span. J Aeronaut Sci 22(3):197–200
22. Lock R (1951) The velocity distribution in the laminar boundary layer between parallel streams. Q J Mech Appl Math 4(1):42–63
23. Lueptow RM (1995) Transducer resolution and the turbulent wall pressure spectrum. J Acoust Soc Am 97(1):370–378. https://doi.org/10.1121/1.412322, http://asa.scitation.org/doi/10.1121/1.412322
24. Michalke A (1972) The instability of free shear layers. Prog Aerosp Sci 12:213–216
25. Miller TS, Gallman JM, Moeller MJ (2012) Review of turbulent boundary-layer models for acoustic analysis. J Aircr 49(6):1739–1754. https://doi.org/10.2514/1.C031405, http://arc.aiaa.org
26. Morkovin MV (1985) Bypass transition to turbulence and research desiderata. In: Conference Proceedings NASA-CP-2386, NASA
27. Peltier L, Hambric S (2007) Estimating turbulent-boundary-layer wall-pressure spectra from CFD RANS solutions. J Fluids Struct 23(6):920–937. https://doi.org/10.1016/J.JFLUIDSTRUCTS.2007.01.003, https://www.sciencedirect.com/science/article/pii/S0889974607000126
28. Pope SB (2006) Turbulent flows, 4th edn. Cambridge University Press, Cambridge
29. Richardson LF (2007) Weather prediction by numerical process. Cambridge University Press, Cambridge
30. Rozenberg Y, Robert G, Moreau S (2012) Wall-pressure spectral model including the adverse pressure gradient effects. AIAA J 50(10):2168–2179
31. Schetz JA (2010) Boundary layer analysis revised. J Fluids Struct
32. Schlichting H, Gersten K (2000) Boundary-layer theory, 8th edn. Springer, Berlin
33. Skåre PE, Krogstad På (????) A turbulent equilibrium boundary layer near separation. J Fluid Mech 319–348. https://doi.org/10.1017/S0022112094004489
34. Slama M, Leblond C, Sagaut P (????) A Kriging-based elliptic extended anisotropic model for the turbulent boundary layer wall pressure spectrum. J Fluid Mech 25–55. https://doi.org/10.1017/jfm.2017.810
35. Tennekes H, Lumley JL, Lumley J, et al (1972) A first course in turbulence. MIT press
36. Wilcox DC et al (1998) Turbulence modeling for CFD, vol 2. DCW industries La Canada, CA

Chapter 7
Flow Noise Generation

Abstract This chapter presents the theory that describes how flow creates sound. Background theory is presented first, which includes Green's functions, multipole sources, a definition of acoustic compactness and a description of the acoustic far field. Lighthill's equation is derived. A solution to Lighthill's equation is used to show that turbulence creates sound like a distribution of quadrupoles. Noise from turbulence, forces and unsteady mass injection is discussed in terms of theory, dimensional analysis and physical interpretation. Curle's theory is presented to develop our understanding of how solid bodies immersed in a flow create sound. The special case of a cylinder placed in a uniform flow is presented and an acoustic model is developed. This model is used with empirical models for flow parameters to illustrate the effect of Reynolds number on cylinder noise emission. The chapter concludes with a presentation of the Ffowcs-Williams and Hawkings equation.

7.1 Preliminary Theory

7.1.1 *Green's Function and the Pressure Solution*

The three-dimensional free-field Green's function was introduced as Eq. 5.54. The frequency domain Green's function is [9],

$$G(\mathbf{x}, \mathbf{y}; \omega) = \frac{-e^{k_0|\mathbf{x}-\mathbf{y}|}}{4\pi|\mathbf{x}-\mathbf{y}|} \tag{7.1}$$

where $\mathbf{x}$ is the position of the observer, $\mathbf{y}$ is the position of the source, $\omega = 2\pi f$ is the rotational frequency, $k_0 = \frac{\omega}{c_0}$ is the wave number, f is frequency and c_0 is the speed of sound in the acoustic medium. We could write $r = |\mathbf{x} - \mathbf{y}|$. Equation 7.1 is the solution to the Helmholtz equation [9], a frequency domain version of the wave equation that relies on the velocity potential.

C. Doolan and D. Moreau, *Flow Noise*,
https://doi.org/10.1007/978-981-19-2484-2_7

For our purposes, we require a time-domain Green's function for a time-domain solution for acoustic pressure. This is [9],

$$G(\mathbf{x}, \mathbf{y}; t) = \frac{1}{4\pi |\mathbf{x} - \mathbf{y}|} \delta\left(t - \tau - |\mathbf{x} - \mathbf{y}|/c_0\right) \tag{7.2}$$

where $\delta\left(t - \tau - |\mathbf{x} - \mathbf{y}|/c_0\right)$ represents an impulsive source that is zero for all time except at $t = \tau$. This Green's function describes a spherical wave created at source position $\mathbf{y}$ travelling outwards with time.

Using the fact that $\rho' = p'/c_0^2$, the general inhomogeneous wave equation with pressure source $\mathcal{F}(\mathbf{y}, t)$ can be written,

$$\frac{1}{c_0^2} \frac{\partial^2 p'}{\partial t^2} - \frac{\partial^2 p'}{\partial x_i^2} = \mathcal{F}(\mathbf{y}, t) \tag{7.3}$$

The solution is obtained by using the Green's function [9],

$$p'(\mathbf{x}, t) = \int \int_{-\infty}^{\infty} \mathcal{F}(\mathbf{y}, \tau) G(\mathbf{x}, \mathbf{y}; t - \tau)\, \mathrm{d}^3\mathbf{y}\, \mathrm{d}\tau \tag{7.4}$$

This equation can be rewritten so that the second integral is performed at *retarded time*,

$$\tau = \tau^* = t - |\mathbf{x} - \mathbf{y}|/c_0 \tag{7.5}$$

The solution is therefore a superposition of contributions from sources at various locations (at different $\mathbf{y}$); the retarded time takes into account the different propagation times so that each wave arrives at the observer location $\mathbf{x}$ at precisely the right time. Combining with the Green's function we obtain the solution for acoustic pressure at position $\mathbf{x}$ for pressure sources located at various positions $\mathbf{y}$,

$$p'(\mathbf{x}, t) = \frac{1}{4\pi} \int_{-\infty}^{\infty} \left[\frac{\mathcal{F}(\mathbf{y}, \tau)}{|\mathbf{x} - \mathbf{y}|} \right]_{\tau = \tau^*} \mathrm{d}^3\mathbf{y} \tag{7.6}$$

7.1.2 *Multipole Sources*

To extend the solution described by Eq. 7.6 to multipole sources (such as monopoles, dipoles, quadrupoles, etc), we can write the pressure source $\mathcal{F}$ as [9],

$$\mathcal{F} = \frac{\partial^n F_{ijk\ldots}}{\partial x_i \partial x_j \partial x_k \cdots} \tag{7.7}$$

which is called a multipole source of order 2^n. Using this, the pressure in the farfield can be determined using a Green's function [9],

$$p'(\mathbf{x}, t) = \frac{\partial^n}{\partial x_i \partial x_j \partial x_k \cdots} \int \int_{-\infty}^{\infty} F_{ijk\ldots}(\mathbf{y}, \tau) G(\mathbf{x}, \mathbf{y}, t - \tau)\, \mathrm{d}^3\mathbf{y}\, \mathrm{d}\tau \tag{7.8}$$

Combining with the three-dimensional free-field Green's function (Eq. 7.2) and performing the integral at retarded time,

$$p'(\mathbf{x}, t) = \frac{\partial^n}{\partial x_i \partial x_j \partial x_k \cdots} \int_{-\infty}^{\infty} \left[\frac{F_{ijk\ldots}(\mathbf{y}, \tau)}{4\pi |\mathbf{x} - \mathbf{y}|} \right]_{\tau=\tau^*} \mathrm{d}^3\mathbf{y} \tag{7.9}$$

This is general solution for the acoustic pressure for a multipole source of order 2^n.

7.1.3 Compact Multipole Sources

A source is said to be *compact* when its characteristic length scale (or its size, which we denote as ℓ) is much smaller than the wavelength of sound it emits (λ). Thus we define a compact source if,

$$k_0 \ell \ll 1 \tag{7.10}$$

where the wavenumber $k_0 = \omega / c_0$.

If the source is at position $\mathbf{y}$ and an observer at position $\mathbf{x}$ so that $r = |\mathbf{x} - \mathbf{y}|$, the observer is said to be in the *acoustic far field* when,

$$k_0 r \gg 1 \tag{7.11}$$

This tells us the observer in the acoustic far field is many wavelengths away from the source and the wave fronts are approximately plane.

Using the chain rule, we can relate spatial and temporal derivatives in retarded time using [8],

$$\frac{\partial f(\tau^*)}{\partial x_i} = \frac{\partial \tau^*}{\partial x_i} \left[\frac{f(\tau)}{\partial \tau} \right]_{\tau=\tau^*} \tag{7.12}$$

where f is a general function. Given the definition for retarded time (Eq. 7.5), we can expand this to [8],

$$\frac{\partial f(\tau^*)}{\partial x_i} = \frac{-(x_i - y_i)}{|\mathbf{x} - \mathbf{y}| c_0} \left[\frac{f(\tau)}{\partial \tau} \right]_{\tau=\tau^*} \tag{7.13}$$

A solution can be found for the acoustic pressure in the far field generated by a compact multipole source by using Eq. 7.13 and ignoring all terms that decay faster than $1/r$ (that is, drop the $1/r^n$ terms for $n > 1$). This expression is [9],

$$p'(\mathbf{x}, t) \approx \frac{(-1)^n x_i x_j x_k \cdots}{4\pi c_0^n |\mathbf{x} - \mathbf{y}|^{n+1}} \frac{\partial^n}{\partial t^n} \int_{-\infty}^{\infty} \left[F_{ijk\ldots}(\mathbf{y}, \tau)\right]_{\tau=\tau^*} \mathrm{d}^3\mathbf{y} \tag{7.14}$$

Equation 7.14 is interesting because it is now cast in terms of a temporal derivative rather than a spatial one, which is sometimes easier to compute. Further, if we set $r = |\mathbf{x} - \mathbf{y}|$, we can see p' depends on both $1/r$ (normal inverse decay law for a spherical acoustic wave) plus the *directional cosines*, $x_i/r, x_j/r, x_k/r, \ldots$.

7.2 Noise from Turbulent Flow

7.2.1 *Lighthill's Equation*

Lighthill [10] proposed a model for the production of *aerodynamic sound* that forms the foundation of flow noise theory today. His insight was that a region of turbulence acts on a surrounding quiescent acoustic medium in the same manner as quadrupole acoustic sources. Lighthill reasoned that turbulence has a fluctuating pressure field whose strength could be reasonably estimated by the fluctuating Reynolds stresses (not the Reynolds stresses normally associated with turbulence that are based on time-averaged quantities). The surrounding acoustic medium (air or water at rest) has fluctuating density and pressure waves that are driven by the fluctuating pressure (or stresses) within the turbulent region. Lighthill's *analogy* describes this concept and this idea rests on the assumption that the turbulent field creates sound but this sound does affect its production.

Turbulent stresses in the flow are balanced by a near-field non-propagating pressure field and an acoustic field, as determined by the governing equations of compressible flow. The near-field pressure remains tied to the turbulence; however, the acoustic field propagates away from the turbulent region, taking with it a tiny amount of energy that we call aerodynamic sound, or flow noise. These concepts will be expanded on first mathematically, before returning to a physical model that can help with its interpretation.

Consider the conservation of mass,

$$\frac{\partial \rho}{\partial t} + \frac{\partial}{\partial x_i}(\rho v_i) = 0 \tag{7.15}$$

where ρ is fluid density, v_i if the flow velocity at position x_i. and time t.

Similarly, the conservation of momentum is,

$$\frac{\partial}{\partial t}(\rho v_i) + \frac{\partial}{\partial x_j}(p_{ij} + \rho v_i v_j) = 0 \tag{7.16}$$

In Lighthill's formulation, The compressive stress tensor is formed using the gauge pressure,

$$p_{ij} = (p - p_0)\delta_{ij} - \sigma_{ij} = p'\delta_{ij} - \sigma_{ij} \tag{7.17}$$

where p is the pressure, p_0 is the pressure of the surrounding acoustic medium, $p' = (p - p_0)$ is the fluctuating pressure and δ_{ij} is the Kronecker delta.[1] The viscous stress tensor is,

$$\sigma_{ij} = \mu\left(\frac{\partial v_i}{\partial x_j} + \frac{\partial v_j}{\partial x_i}\right) - \frac{2}{3}\mu\left(\frac{\partial v_k}{\partial x_k}\right)\delta_{ij} \tag{7.18}$$

where μ is the dynamic viscosity of the fluid.

Taking the time derivation of Eq. 7.15,

$$\frac{\partial^2}{\partial t \partial x_i}(\rho v_i) = -\frac{\partial^2 \rho}{\partial t^2} \tag{7.19}$$

and taking the divergence (spatial derivative) of Eq. 7.16,

$$\frac{\partial^2}{\partial t \partial x_i}(\rho v_i) = -\frac{\partial^2}{\partial x_i \partial x_j}(p_{ij} + \rho v_i v_j) \tag{7.20}$$

Subtract Eq. 7.20 from Eq. 7.19, we obtain:

$$\frac{\partial^2 \rho}{\partial t^2} = \frac{\partial^2}{\partial x_i \partial x_j}(p_{ij} + \rho v_i v_j) \tag{7.21}$$

Using the fact that the derivative of density is the same as the derivative of density fluctuation, that is,

$$\frac{\partial \rho}{\partial t} = \frac{\partial(\rho - \rho_0)}{\partial t} = \frac{\partial \rho'}{\partial t} \tag{7.22}$$

where ρ_0 is the density of the surrounding acoustic medium and $\rho' = (\rho - \rho_0)$ is the fluctuating density.

[1] $\delta_{ij} = 1$ if $i = j$, $\delta_{ij} = 0$ if $i \neq j$.

Using this, we introduce the term,

$$c_0^2 \frac{\partial^2 \rho'}{\partial x_i^2} = \frac{\partial^2 c_0^2 \rho' \delta_{ij}}{\partial x_i^2} \tag{7.23}$$

Subtracting Eq. 7.23 from both sides of Eq. 7.21, we obtain Lighthill's famous equation,

$$\frac{\partial^2 \rho'}{\partial t^2} - c_0^2 \frac{\partial^2 \rho'}{\partial x_i^2} = \frac{\partial^2 T_{ij}}{\partial x_i \partial x_j} \tag{7.24}$$

where T_{ij} is Lighthill's stress tensor,

$$T_{ij} = \rho v_i v_j + (p' - c_0^2 \rho')\delta_{ij} - \sigma_{ij} \tag{7.25}$$

Equation 7.24 is a non-homogeneous wave equation that shows aerodynamic sound is produced by a distribution of quadrupoles [10]. It assumes these sources are within a turbulent region and act on an unbounded, surrounding acoustic medium. No assumptions are made in the derivation of Lighthill's equation, also known as Lighthill's analogy. Hence it is no easier to solve than the original governing equations. What it does provide is insight into the sound producing physics.

The strength of the source quadrupoles is determined by Lighthill's tensor, T_{ij}, Eq. 7.25 and consists of three terms that describe different processes that contribute to the quadrupole strength. These are the fluctuating Reynolds stresses, $\rho v_i v_j$, an "isentropic deviation term", $(p' - c^2\rho')\delta_{ij}$ and a viscous term, σ_{ij}.

The fluctuating Reynolds stress term is non-linear and relates turbulent motion as unsteady stress (pressure), which in turn induces sound on the surrounding medium. The isentropic deviation term compares the pressure fluctuation with what would be expected in an equivalent "linear" acoustic field that has the same density fluctuation. If there are non-linearities from entropy variations, then this term will not be zero and will contribute to sound production. Finally, the viscous term accounts for attenuation of sound. This term is normally ignored because at high Reynolds number, usually the case for engineering flows, its value is small in the source region. Outside the source zone, where the motion is linear (the acoustic radiation field) the observer locations are normally reasonably close to the source region making viscous attenuation negligible. Note that outside the source region, $T_{ij} = 0$, and Eq. 7.24 reverts to a classical homogeneous wave equation.

For low Mach number flows that are homoentropic and high Reynolds number, Lighthill's stress term can be approximated as [10],

$$T_{ij} \approx \rho_0 v_i v_j \tag{7.26}$$

This is because the error associated with this approximation is of the order $O(M^2)$ [10], thus is low when $M \leq 0.3$. Intuitively, we can see that if homoentropic, the isentropic deviation term should be zero ($p' - c_0^2\rho' \approx 0$) and the density fluctuation much lower than the ambient density ($\rho' \ll \rho_0$). Hence, the approximation is reasonable.

7.2.2 Solution to Lighthill's Equation

We can apply the general multipole solution of Eq. 7.9 to Lighthill's equation. For a quadrupole source ($n = 2$, order $2^n = 4$) so that,

$$\mathcal{F}(\mathbf{y}, t) = \frac{\partial^2 T_{ij}(\mathbf{y}, t)}{\partial x_i \partial x_j} \tag{7.27}$$

and

$$F_{ij}(\mathbf{y}, t) = T_{ij}(\mathbf{y}, t) \tag{7.28}$$

We can now see that the far-field acoustic pressure solution is,

$$p'(\mathbf{x}, t) = \frac{\partial^2}{\partial x_i \partial x_j} \int_{-\infty}^{\infty} \left[\frac{T_{ij}(\mathbf{y}, \tau)}{4\pi |\mathbf{x} - \mathbf{y}|} \right]_{\tau=\tau^*} \mathrm{d}^3\mathbf{y} \tag{7.29}$$

If we assume a compact source, the acoustic pressure in the far field can be approximated by,

$$p'(\mathbf{x}, t) \approx \frac{x_i x_j}{4\pi c_0^2 |\mathbf{x} - \mathbf{y}|^3} \frac{\partial^2}{\partial t^2} \int_{-\infty}^{\infty} \left[T_{ij}(\mathbf{y}, \tau) \right]_{\tau=\tau^*} \mathrm{d}^3\mathbf{y} \tag{7.30}$$

7.2.3 Physical Interpretation

The right-hand-side of Eq. 7.24 shows that turbulence creates sound just like a distribution of quadrupoles exerting a stress on a surrounding acoustic medium. A quadrupole is formed by placing two equal but opposite dipoles adjacent to each other, and taking the limit of their separation distance to zero (Sect. 5.7.2). By arranging the dipoles in various ways *longitudinal* and *lateral* quadrupoles can be formed. These quadrupoles were shown earlier in Figures 5.9 and 5.10. If the dipole represents a force acting on acoustic medium (Sect. 5.7.2), then the quadrupole represents two opposing forces, otherwise known as a *stress*.

A pressure stress therefore can be represented by a longitudinal quadrupole. Similarly, a shear stress can be represented by the lateral quadrupole.

A useful physical interpretation can be found if we consider a single, compact eddy so that Lighthill's stress term can be approximated by Eq. 7.26. Equation 7.30 shows the far field pressure is determined by the product of the two directional cosines. Considering the (x_1, x_2) plane, θ is the angle the vector $\mathbf{x} - \mathbf{y}$ makes with the x_1 axis. We can then show that the products of the directional cosines are,

$$\frac{x_1}{r}\frac{x_1}{r} = \frac{\cos^2\theta}{r^2} \tag{7.31}$$

$$\frac{x_1}{r}\frac{x_2}{r} = \frac{\cos\theta\sin\theta}{r^2} \tag{7.32}$$

$$\frac{x_2}{r}\frac{x_2}{r} = \frac{\sin^2\theta}{r^2} \tag{7.33}$$

where $r = |\mathbf{x} - \mathbf{y}|$. Thus two longitudinal quadrupoles (Eqs. 7.31 and 7.33) and one lateral quadrupole (Eq. 7.32) are formed. Considering the approximation of T_{ij} given by Eq. 7.26, we can see that longitudinal quadrupoles have strengths given by $T_{11} = \rho_0 v_1 v_1$ and $T_{22} = \rho_0 v_2 v_2$. Further, we can link the fluctuating normal Reynolds stresses $(v_1 v_1, v_2 v_2)$ directly to a particular types of quadrupoles with their own characteristic directivity patterns.

Following on, a lateral quadrupole is created with strength $T_{12} = \rho_0 v_1 v_2$. Here, Eq. 7.32 shows that the directivity of this eddy is the same as a lateral quadrupole. This links the fluctuating Reynolds shear stress to lateral quadrupoles. Note that this illustrative example limits the discussion to a single plane. In three dimensions we have three longitudinal quadrupoles and six lateral quadrupoles corresponding to the normal and shear fluctuating Reynolds stresses that comprise Lighthill's stress tensor.

7.2.4 Dimensional Analysis

A dimensional analysis can be performed to obtain an order of magnitude estimate of acoustic pressure in the far field generated by turbulent eddies. A dimensional analysis also provides a velocity scaling law.

We can consider turbulent flow as a collection of uncorrelated eddies with characteristic length scale ℓ. The velocity scale associated with the eddy is U_c, the convective velocity of the eddy [1]. We now can assume that Lighthill's stress tensor can be approximated by an order of magnitude estimate,

$$T_{ij} \approx \rho_0 U_c^2 \tag{7.34}$$

Further, we can assume that the time derivative can be approximated by the characteristic frequency,

$$\frac{\partial}{\partial t} \approx \frac{U_c}{\ell} \tag{7.35}$$

which can be thought of as an average eddy turn-over rate.

If we assume that each eddy is acoustically compact with volume $V \approx \ell^3$, we can use Eq. 7.30 to give an order of magnitude estimate of far field pressure,

$$p' \approx \frac{1}{4\pi c^2 r} \frac{U_c^2}{\ell^2} \rho_0 U_c^2 \ell^3 \approx \frac{\ell}{r} \rho_0 U_c^2 M_c^2 \tag{7.36}$$

where the 4π term has been dropped for brevity.

Recalling that acoustic intensity is (Sect. 5.6),

$$I = p'u = \frac{p'^2}{\rho_0 c_0} \tag{7.37}$$

Thus, an order of magnitude estimate of acoustic intensity for a compact eddy of turbulence is,

$$I \approx \rho_0 \frac{\ell^2}{r^2} U_c^3 M_c^5 \tag{7.38}$$

which is Lighthill's famous eighth power law, which illustrates the very high sensitivity of radiated sound intensity to the speed with which the eddy travels. From dimensional analysis, an estimate of the energy supplied to the flow [10] (N m s^{-1}, that is also removed by turbulent decay),

$$\mathcal{E} \approx \rho_0 U_c^3 \ell^2 \tag{7.39}$$

Thus the efficiency of aerodynamic sound production can be estimated as,

$$\eta = \frac{IV}{\mathcal{E}} \propto M_c^5 \tag{7.40}$$

Hence, we see that not only is aerodynamic sound level very sensitive to flow velocity, it is a only a minute fraction of the steady flow energy. For subsonic flows where $M_c \ll 1$, it is a tiny amount indeed. This indicates the inefficiency of turbulence as a source of sound. It also explains why incompressible flow simulations can be used with an acoustic analogy to predict sound (a compressible phenomena) because the amount of energy escaping the flow as acoustic energy is so low it does not affect the flow itself.

7.3 Noise from Unsteady Mass Injection

7.3.1 Theory

Consider the situation where fluid is injected into a quiescent fluid. If this mass source is $\dot{m}$ (units kg m^{-3} s^{-1}), then we can write the conservation of mass as,

$$\frac{\partial \rho}{\partial t} + \frac{\partial}{\partial x_i}(\rho v_i) = \dot{m} \tag{7.41}$$

The surrounding fluid is quiescent and we also assume to be invscid. We can write the momentum equation as,

$$\frac{\partial}{\partial t}(\rho v_i) + \frac{\partial}{\partial x_j}(p_{ij}) = 0 \tag{7.42}$$

where we can assume $p_{ij} = p'\delta_{ij}$.

Taking the temporal derivative of Eq. 7.41 and rearranging,

$$\frac{\partial^2}{\partial t \partial x_i}(\rho v_i) = \frac{\partial}{\partial t}(\dot{m}) - \frac{\partial^2 \rho}{\partial t^2} \tag{7.43}$$

Similarly, taking the spatial derivative of the momentum equation (Eq. 7.42) and rearrange,

$$\frac{\partial^2}{\partial t \partial x_i}(\rho v_i) = -\frac{\partial^2}{\partial x_i x_j}(p'\delta_{ij}) = -\nabla^2 p' \tag{7.44}$$

Equating Eqs. 7.43 and 7.44, and using the homoentropic equation of state $p' = \rho'/c_0^2$,

$$\frac{1}{c_0^2}\frac{\partial^2 p'}{\partial t^2} - \nabla^2 p' = \frac{\partial}{\partial t}(\dot{m}) \tag{7.45}$$

This leaves us with the wave equation in pressure form, with a monopole source consisting of the fluctuating mass flow rate of the injecting gas. We see that steady mass injection creates no sound.

Note that the source mass flow rate may be considered as the displaced mass flux in the surrounding medium. We can then write,

$$\dot{m} = \rho_0 q \tag{7.46}$$

where $q(\mathbf{x}, t)$ is the instantaneous volumetric flow rate. Now, the wave equation (Eq. 7.45) can be written,

$$\frac{1}{c_0^2}\frac{\partial^2 p'}{\partial t^2} - \nabla^2 p' = \rho_0 \frac{\partial q}{\partial t} \tag{7.47}$$

7.3.2 Solution

We write the solution for a compact monopole mass source. A compact monopole source is a multipole of order 1 ($n = 0$) so that,

$$\mathcal{F}(\mathbf{y}, t) = \frac{\partial q}{\partial t} \tag{7.48}$$

Noting that $\mathcal{F} = F$ for the monopole case. Using the general solution (Eq. 7.14) to solve for p',

$$p'(\mathbf{x}, t) = \frac{\rho_0}{4\pi |\mathbf{x} - \mathbf{y}|} \int_{-\infty}^{\infty} \left[\frac{\partial q}{\partial t} \right]_{\tau=\tau^*} \mathrm{d}^3\mathbf{y} \tag{7.49}$$

The total volume flow rate of the source is,

$$\frac{\partial V}{\partial t} = \dot{V} = \int_{-\infty}^{\infty} q(\mathbf{y}, t)\, \mathrm{d}^3\mathbf{y} \tag{7.50}$$

Hence, the pressure field is given by,

$$p'(\mathbf{x}, t) = \frac{\rho_0 \ddot{V}(t)_{\tau=\tau^*}}{4\pi |\mathbf{x} - \mathbf{y}|} \tag{7.51}$$

This tells us it is the acceleration of the volume displacement which controls noise level in the far field.

7.3.3 Physical Interpretation

The compact volumetric source can represent multiple flow noise sources. The first is cavitation in liquid flows. Cavitation occurs when the local pressure is less that a certain critical pressure where bubbles form [2]. These bubble can grow and collapse, thus mimicking the volume source of Eq. 7.51 and radiate sound.

Bubbles in turbulent flows and jets may also emit sound like a compact monopole [1, 2]. Similar to the cavitation case, the turbulent stresses within the jet can force the bubbles to oscillates and create sound.

Another important source is due to combustion [6]. Consider a small region (compared with wavelength so it is compact) of perfectly mixed fuel and air surrounded by a quiescent gas. If this fuel-air mixture ignites, it will expand. The expansion of this region will create sound as it accelerates the surrounding medium, as described in the model of Eq. 7.51.

7.3.4 Dimensional Analysis

Consider a jet orifice diameter of d that creates an oscillating mass source of sound. The velocity scale associated with the jet is U_j. The first and second time derivatives can therefore be written as,

$$\frac{\partial}{\partial t} \sim \frac{U_j}{d} \qquad \frac{\partial^2}{\partial t^2} \sim \frac{U_j^2}{d^2} \tag{7.52}$$

Using these approximations,

$$p' \sim \frac{\rho_0 U_j^2}{d^2 r} \tag{7.53}$$

The intensity can be therefore be estimated,

$$I = p'u = \frac{p'^2}{\rho_0 c_0} \approx \frac{\rho_0 U_j^4}{c_0 d^4 r^2} = \frac{\rho_0 M_j U_j^3}{d^4 r^2} \tag{7.54}$$

Hence, we see that the compact oscillating mass source has an acoustic power dependence proportional to the fourth power of the characteristic jet velocity. Compared with turbulence, it is a much more efficient radiator of sound.

7.4 Noise from Unsteady Forces

7.4.1 Theory

We now turn our attention to the case of a compact unsteady force applied to a flow at point $\mathbf{y}$. If this force can be written as -$\mathbf{f}(\mathbf{y}, t)$ (units N m^{-3}), we can write the mass and momentum conservation equations as,

$$\frac{\partial \rho}{\partial t} + \frac{\partial}{\partial x_i}(\rho v_i) = 0 \tag{7.55}$$

$$\frac{\partial}{\partial t}(\rho v_i) + \frac{\partial}{\partial x_i}(p_{ij}) = -f_i(\mathbf{y}, t) \tag{7.56}$$

where we assume an inviscid fluid as in the previous section. Taking the temporal derivative of Eq. 7.55 and the spatial derivative of Eq. 7.56 then equating, leaves us with,

$$\frac{1}{c_0^2}\frac{\partial^2 p'}{\partial t^2} - \nabla^2 p' = -\frac{\partial f_i(\mathbf{y}, t)}{\partial x_i} \tag{7.57}$$

This is a non-homogeneous wave equation in pressure form with a dipole source term equal to the spatial gradient of force.

7.4.2 Solution

Assuming the force is compact, the source is multipole of order 2 ($n = 1$) so that,

$$\mathcal{F}(\mathbf{y}, t) = -\frac{\partial f_i(\mathbf{y}, t)}{\partial x_i} \tag{7.58}$$

Hence, following Eq. 7.14, we can write the solution for pressure p',

$$p'(\mathbf{x}, t) \approx \frac{x_i}{4\pi c_0 |\mathbf{x} - \mathbf{y}|^2} \frac{\partial f_i(\mathbf{y}, t)}{\partial t} \tag{7.59}$$

7.4.3 Physical Interpretation

The compact force source is used to represent the case of acoustically small objects placed in a flow. The unsteady aerodynamic forces generated during this process supports an acoustic dipole. As you can imagine, this is an extremely important topic in flow noise theory and is discussed in more detail in Sect. 7.5 which describes Curle's theory. For now, it is important to notice that the acoustic field is controlled by the product of the directional cosines and the temporal derivative of the component of force aligned with it. In a two-dimensional (x_1, x_2) plane these are,

$$\frac{x_1}{r} \times \frac{\partial f_1}{\partial t} = \dot{f}_1 \frac{\cos\theta}{r} \tag{7.60}$$

$$\frac{x_2}{r} \times \frac{\partial f_2}{\partial t} = \dot{f}_2 \frac{\sin\theta}{r} \tag{7.61}$$

where $\mathbf{r} = \mathbf{x} - \mathbf{y}$ and θ is the angle between $\mathbf{r}$ and the x_1 axis. Thus the acoustic field can be described as two lobes aligned with the directions of fluctuating force ($\pm\mathbf{f}$).

7.4.4 Dimensional Analysis

Consider an object of characteristic dimension d (the diameter of a sphere, for example) placed in a flow with velocity U. As previously written, the time derivative can be written,

$$\frac{\partial}{\partial t} \sim \frac{U}{d} \tag{7.62}$$

We can also approximate the force term f_i as

$$f_i \approx \frac{\rho_\infty U^2}{d} \tag{7.63}$$

Substituting into Eq. 7.59, we obtain,

$$p' \approx \frac{1}{4\pi c_0 d r} \rho_\infty U^3 \tag{7.64}$$

where we assume $x_i/r \approx 1$. This means we are estimating the pressure at the peak location in the dipole pressure lobe.

The intensity is estimated as,

$$I = p'u = \frac{p'^2}{\rho_\infty c_0} \sim \frac{\rho_\infty M^3 U^3}{d^2 r^2} \tag{7.65}$$

Thus, we see that the compact fluctuating force has an acoustic power proportional to the sixth power of velocity. This type of force has an efficiency that is in-between the weak quadrupole source of turbulence and relatively strong monopole source that represents unsteady volume or mass injection.

7.5 Curle's Theory

We now consider the case where there are solid boundaries within the turbulent flow field. The typical case is where turbulence is shed from an object (wing, bluff body, etc) in uniform flow. Another case is where turbulence within the flow, usually approaching from upstream, interacts with a solid body (a wing, for example). In these cases, the solid body scatters (or in some cases where the wavelength is small, diffracts) the sound created by the turbulence field. The scattered waves change the nature of the sound sources. Curle [4] extended Lighthill's theory to accommodate solid boundaries, known as the Theorem of Curle, or more simply, Curle's theory. This section will present a derivation of Curle's theory plus an application to the case of a cylinder in cross flow.

7.5.1 General Derivation

Consider the case where a region of turbulent flow also contains a solid surface. The effect of the solid surface is to reflect and diffract sound from the quadrupoles we use to model aerodynamic sound. To take into account the effect of this solid surface on sound production and radiation, Curle [4] uses Kirchoff's equation (attributed to Stratton [12] by Curle [4] and also derived by Blake [1]) to obtain a solution to Lighthill's equation (Eq. 7.24). We write this solution as,

$$p'(\mathbf{x},t) = \frac{1}{4\pi}\int_V \left[\frac{\partial^2 T_{ij}}{\partial y_i \partial y_j}\right]_{\tau=\tau^*} \frac{\mathrm{d}V(\mathbf{y})}{r} + \frac{c_0^2}{4\pi}\int_S \left[\frac{1}{r}\frac{\partial \rho}{\partial n} + \frac{1}{r^2} + \frac{1}{c_0 r}\frac{\partial r}{\partial n}\frac{\partial \rho}{\partial t}\right]_{\tau=\tau^*} \mathrm{d}S(\mathbf{y}) \tag{7.66}$$

Here, we have shortened the solution by assuming that the volume of fluid external to the surface (containing the turbulent flow and quiescent media) is volume V and the surface is represented by area S. Note that we have shorted some of the functions, so that $T_{ij} = T_{ij}(\mathbf{y}, \tau)$ and $\rho = \rho(\mathbf{y}, \tau)$. Also recall $r = |\mathbf{x} - \mathbf{y}|$ and $\mathbf{n}$ is the surface outward normal from the fluid,

$$\mathbf{n} = l_i \tag{7.67}$$

Curle [4] shows that by applying the divergence theorem twice, the volume integral for the turbulent source term can be written,

$$\int_V \left[\frac{\partial^2 T_{ij}}{\partial y_i \partial y_j}\right]_{\tau=\tau^*} \frac{\mathrm{d}V(\mathbf{y})}{r} = \frac{\partial^2}{\partial x_i \partial x_j}\int_V \left[T_{ij}\right]_{\tau=\tau^*} \frac{\mathrm{d}V(\mathbf{y})}{r} + \frac{\partial}{\partial x_i}\int_S l_j \left[T_{ij}\right]_{\tau=\tau^*} \frac{\mathrm{d}S(\mathbf{y})}{r} + \int_S l_i \left[\frac{\partial T_{ij}}{\partial y_j}\right]_{\tau=\tau^*} \frac{\mathrm{d}S(\mathbf{y})}{r} \tag{7.68}$$

Using the identity,

$$\frac{\partial}{\partial x_i}\left\{\frac{1}{r}[f]_{\tau=\tau^*}\right\} = -\left\{\frac{1}{r^2}f + \frac{1}{c_0 r}f'\right\}\frac{\partial r}{\partial x_i} \tag{7.69}$$

Curle shows the surface integral term can be written,

$$\int_S \left[\frac{1}{r}\frac{\partial \rho}{\partial n} + \frac{1}{r^2} + \frac{1}{c_0 r}\frac{\partial r}{\partial n}\frac{\partial \rho}{\partial t}\right]_{\tau=\tau^*} \mathrm{d}S(\mathbf{y}) = \int_S l_i \frac{1}{r}\frac{\partial}{\partial y_j}(\rho\delta_{ij})\mathrm{d}S(\mathbf{y}) + \int_S l_j \frac{\partial}{\partial x_i}\left(\frac{1}{r}\rho\delta_{ij}\right)\mathrm{d}S(\mathbf{y}) \tag{7.70}$$

Substituting Eqs. 7.68 and 7.70 into Eq. 7.66 we obtain,

$$p'(\mathbf{x},t) = \frac{1}{4\pi}\frac{\partial^2}{\partial x_i \partial x_j}\int_V \left[T_{ij}\right]_{\tau=\tau^*}\frac{\mathrm{d}V(\mathbf{y})}{r} + \frac{1}{4\pi}\int_S l_i \frac{1}{r}\frac{\partial}{\partial y_j}\left(T_{ij} + c_0^2\rho\delta_{ij}\right)\mathrm{d}S(\mathbf{y}) + \frac{1}{4\pi}\frac{\partial}{\partial x_i}\int_S l_j \frac{1}{r}\left(T_{ij} + c_0^2\rho\delta_{ij}\right)\mathrm{d}S(\mathbf{y}) \tag{7.71}$$

Rewriting Lighthill's stress tensor (Eq. 7.25) to include the compressive stress tensor (Eq. 7.17),

$$T_{ij} = \rho v_i v_j + p_{ij} - c_0^2\rho\delta_{ij} \tag{7.72}$$

Substituting Eq. 7.72 into Eq. 7.71 we obtain,

$$p'(\mathbf{x},t) = \frac{1}{4\pi}\frac{\partial^2}{\partial x_i \partial x_j}\int_V \left[T_{ij}\right]_{\tau=\tau^*}\frac{\mathrm{d}V(\mathbf{y})}{r} + \frac{1}{4\pi}\int_S l_i \frac{1}{r}\frac{\partial}{\partial y_j}\left(\rho v_i v_j + p_{ij}\right)\mathrm{d}S(\mathbf{y}) + \frac{1}{4\pi}\frac{\partial}{\partial x_i}\int_S l_j \frac{1}{r}\left(\rho v_i v_j + p_{ij}\right)\mathrm{d}S(\mathbf{y}) \tag{7.73}$$

Using the conservation of momentum (Eq. 7.16),

$$l_i \frac{\partial}{\partial y_j}(\rho v_i v_j + p_{ij}) = -l_i \frac{\partial}{\partial t}(\rho v_i) \tag{7.74}$$

and if the surfaces are not moving (or vibrating) then,

$$l_i v_i = 0 \tag{7.75}$$

we can then reduce Eq. 7.73 to,

$$p'(\mathbf{x},t) = \frac{1}{4\pi}\frac{\partial^2}{\partial x_i \partial x_j}\int_V \left[T_{ij}\right]_{\tau=\tau^*}\frac{\mathrm{d}V(\mathbf{y})}{r} + \frac{1}{4\pi}\frac{\partial}{\partial x_i}\int_S \frac{1}{r} l_j p_{ij}\,\mathrm{d}S(\mathbf{y}) \tag{7.76}$$

Using,

$$P_i = -l_j p_{ij} \tag{7.77}$$

we can write,

$$p'(\mathbf{x},t) = \frac{1}{4\pi}\frac{\partial^2}{\partial x_i \partial x_j}\int_V \left[T_{ij}\right]_{\tau=\tau^*}\frac{\mathrm{d}V(\mathbf{y})}{r} - \frac{1}{4\pi}\frac{\partial}{\partial x_i}\int_S \left[P_i\right]_{\tau=\tau^*}\frac{\mathrm{d}S(\mathbf{y})}{r} \tag{7.78}$$

which is known as Curle's theory. Curle's theory states that the sound created by a solid object placed in a turbulent flow is modelled exactly as a volume distribution of quadrupoles (of strength T_{ij}) in the turbulent regions and a surface distribution of dipoles (of strength P_i) on the solid object. The dipole strength P_i is the force per unit area exerted on the fluid by the solid boundaries in the x_i direction [4]. No simplifying assumptions have been made; Eq. 7.78 includes the effects of convection, dissipation and acoustic diffraction. As is the case with Lighthill's equation (Eq. 7.24), the theory is exact; however, obtaining accurate predictions relies upon an accurate estimation of the source terms T_{ij} and P_i.

7.5.2 Low Mach Number Compact Case

The theorem of Curle makes use of both quadrupole and dipole terms. Our previous presentation of quadrupole and dipole sources reveals that they have different radiation efficiencies. Inspecting the approximate relationships for acoustic intensity for quadrupoles and dipoles (Eqs. 7.38 and 7.65) we can form the ratio,

$$\frac{I_q}{I_d} \sim M^2 \tag{7.79}$$

where I_q and I_d are the acoustic intensities for a quadrupole and dipole respectively. This represents the ratio of contributions to an acoustic field by sources created in equivalent flows. For low Mach number flows, where $M \lll 1$, the contribution of the quadrupole terms (that is, the sound created by turbulence) to the acoustic field becomes small. In this case, the dipole terms dominate and Curle's theory reduces to,

$$p'(\mathbf{x}, t) = -\frac{1}{4\pi}\frac{\partial}{\partial x_i}\int_S [P_i]_{\tau=\tau^*}\ \frac{\mathrm{d}S(\mathbf{y})}{r} \tag{7.80}$$

If we assume that the solid surface is acoustically compact and the observer is in the far field (Sect. 7.1.3), we can use Eq. 7.14 to solve Eq. 7.80,

$$p'(\mathbf{x}, t) = \frac{1}{4\pi c_0}\frac{x_i}{r^2}\frac{\partial}{\partial t}[F_i(\tau)]_{\tau=\tau^*} \tag{7.81}$$

where,

$$F_i(t) = \int_S P_i(\mathbf{y}, t)\ \mathrm{d}S(\mathbf{y}) \tag{7.82}$$

is the resultant force that the solid body exerts on the surrounding fluid.

Equation 7.81 is extremely useful. It is much easier to calculate (or measure) the force on an object than to estimate the pressure at all points on its surface. Using this value of force, low frequency noise level estimates can be made.

Our discussion here also links neatly with the general solution for the acoustic field generated by a compact point source shown in Sect. 7.4.

7.6 Cylinder Noise

The noise created by a cylinder in cross flow is an excellent vehicle to illustrate the physics surrounding noise sources created by solid objects in turbulent flow. Cylinder flow noise is also a very practical example, with cylinders representing the flow over bluff bodies in general, which may appear in aeronautical, maritime and industrial situations.

This section will give a brief overview of the unsteady fluid mechanics of "two-dimensional" cylinders in cross flow. By two-dimensional, we mean that the cross-section of the cylinder is constant along the span. The span is not zero and is normally considered to be finite. the flow over the cylinder, if the Reynolds number is high enough, is not two-dimensional due to the three-dimensional nature of transition and turbulence in the flow about the cylinder and its wake.

7.6.1 Flow over a Cylinder

For most practical purposes, the flow over a cylinder is considered to be unsteady. In fact, above a Reynolds number based on diameter of $Re_d = \frac{U_\infty d}{\nu} \sim 47$ [11, 15] (where d is the cylinder diameter) the wake forms a vortex street, a series of almost periodic vortices behind the cylinder. These vortices are the cause of unsteady lift and drag forces on the cylinder. They are the cause of the so-called Aeolian tone, first observed by Strouhal [13].

For simplicity and for this section only, we will drop the subscript d for Reynolds number so that $Re = Re_d$).

7.6.1.1 Vortex Shedding and the Strouhal Number

Figure 7.1 illustrates the vortex shedding process behind a cylinder. Uniform flow with freestream velocity U_∞ passes over a circular cylinder of diameter d. Provided the Reynolds number is high enough, vortices form in the wake. These are created by the interaction of the separated boundary layers on the cylinder with the low-pressure base flow region (immediately downstream of the cylinder). The shear layers become unstable and "roll-up" into vortices that convect downstream. Figure 7.1 shows a two-dimensional slice of a cylinder flow. The vortices in the wake are transported downstream with mean convection velocity U_c. The vortices have a streamwise separation distance of λ_v. Therefore, we can write a general relationship for the shedding frequency, f_s,

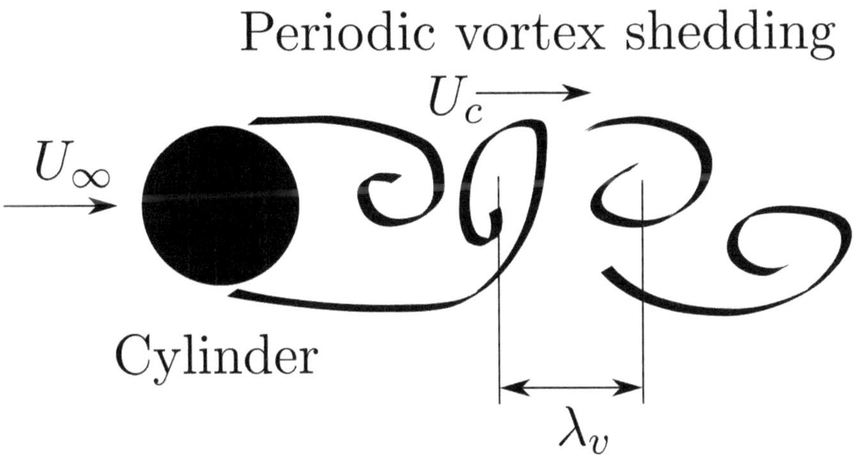

Fig. 7.1 Illustration of unsteady vortex shedding from a circular cylinder placed in a cross flow with freestream velocity U_∞

$$f_s = \frac{U_c}{\lambda_v} \tag{7.83}$$

We normally express the shedding frequency non-dimensionalised by the freestream velocity to form the Strouhal number,

$$St = \frac{f_s d}{U_\infty} \tag{7.84}$$

In fact, we often non-dimensionalise all frequencies this way, thus a Strouhal number can be formed at any frequency, not just the shedding frequency.

At the shedding frequency, the Strouhal number varies with Reynolds number. This is because the wake flow is affected by viscous effects including wake transition and the boundary layer separation point on the cylinder [15]. The various shedding regimes was classified by Williamson [15], which will be summarised below.

For $Re \lesssim 47$, the flow is steady, two-dimensional and there is no vortex shedding. A recirculation region forms behind the cylinder. The length of this recirculation region grows with Reynolds number and this is known as the laminar steady regime.

As the Reynolds number increases ($47 \lesssim Re \lesssim 194$), the recirculation region becomes unstable and forms a wake oscillation that is classified as the laminar-shedding regime. The shedding in this regime is also considered to be mostly, but not exclusively, two-dimensional.

The three-dimensional wake-transition regime occurs when $190 \lesssim Re \lesssim 260$. In this regime, streamwise vorticity and so-called "vortex dislocations" occur, which are also referred to as "mode A" and "mode B" shedding. This is the regime before fine-scale three-dimensional instabilities progressively become more important in the cylinder wake. Over the range $260 \lesssim Re \lesssim 1000$, these "three-dimensionalities" progressively reduce the base-pressure, lengthen the recirculation region and increase the shedding Strouhal number.

The shear-layer transition regime ($1000 \lesssim Re \lesssim 2 \times 10^5$) is characterised by increasing base pressure with Reynolds number with a decrease in formation length and an increase in shedding frequency. The Kelvin Helmholtz instability is excited in the the separated shear layers and is responsible for an increase in Reynolds stress. Three-dimensional structures are also formed and contribute to the development of a complex wake that has a random phase variation in vortex shedding along the span.

Note that cylinder flow in the range $260 \lesssim Re \lesssim 2 \times 10^5$ is also known as the sub-critical regime.

Immediately after the sub-critical regime, the wake flow undergoes a sudden change in a narrow Reynolds number range ($2 \times 10^5 \lesssim Re \lesssim 3.5 \times 10^5$) known as the critical (or critical-transition) regime. Here, drag decreases greatly due to a reduced wake width. Further, asymmetric reattachment of separated shear layers may occur on the cylinder surface causing high oscillating lift forces.

As the Reynolds number increases again ($3.5 \times 10^5 \lesssim Re \lesssim 1.5 \times 10^6$), the flow moves into the supercritical regime. Flow separation and reattachment is symmetric in this case. The shedding Strouhal number increases quite considerably in this region and the boundary layer has become transitional.

Up until this point, the transition sequence has occurred in this order: wake transition ($Re \sim 190 - 260$), shear layer transition ($Re \sim 1000 - 2 \times 10^5$) and the beginnings of boundary layer transition ($Re \sim 2 \times 10^5 - 1.5 \times 10^6$). In the post-critical (or boundary-layer transition) regime ($Re \gtrsim 1.5 \times 10^6$) the boundary layer becomes turbulent on the surface of the cylinder. Once this occurs, the separation point moves aft and the drag decreases. The wake is also highly turbulent. It is interesting to note that coherent vortex shedding can still be observed even in the post-critical shedding regime.

Norberg [11] has collated a number of experimental and numerical results to form an empirical model of the variation of Strouhal number with Reynolds number based on diameter. To cope with with the various shedding regimes, the model has many sub-components and these are presented in Table 7.1. Figure 7.2 shows this empirical model. The various shedding regimes can be seen to affect shedding Strouhal number as Reynolds number is varied.

7.6.1.2 Unsteady Forces

As discussed above, a cylinder in crossflow will shed vortices at a shedding frequency which we can non-dimensionalise into a shedding Strouhal number. The shed vorticity will create unsteady forces on the cylinder.

First, there will be an unsteady lift force that oscillates at the shedding frequency. This is because of the alternate sign of the circulation of successive vortices shed from the cylinder. Each vortex will induce a circulation about the cylinder and hence

Table 7.1 Shedding frequency Strouhal number empirical functions [11]

$Re =$	$St \simeq$	
$47 - 190$	$0.2663 - 1.019/\sqrt{Re}$	
$165 - 260$	$-0.089 + 22.9/Re + 7.8 \times 10^{-4} Re$	
$260 - 325$	0.2016	
$325 - 1.6 \times 10^3$	$0.2139 - 4.0/Re$	
$1.6 \times 10^3 - 1.5 \times 10^5$	$0.1853 + 0.0261 \times \exp(-0.9 \times x^{2.3})$	$x = \log(Re/1.6 \times 10^3)$
$1.5 \times 10^5 - 3.4 \times 10^5$	$0.1848 + 8.6 \times 10^{-4} \times (Re/1.5 \times 10^5)^{4.6}$	

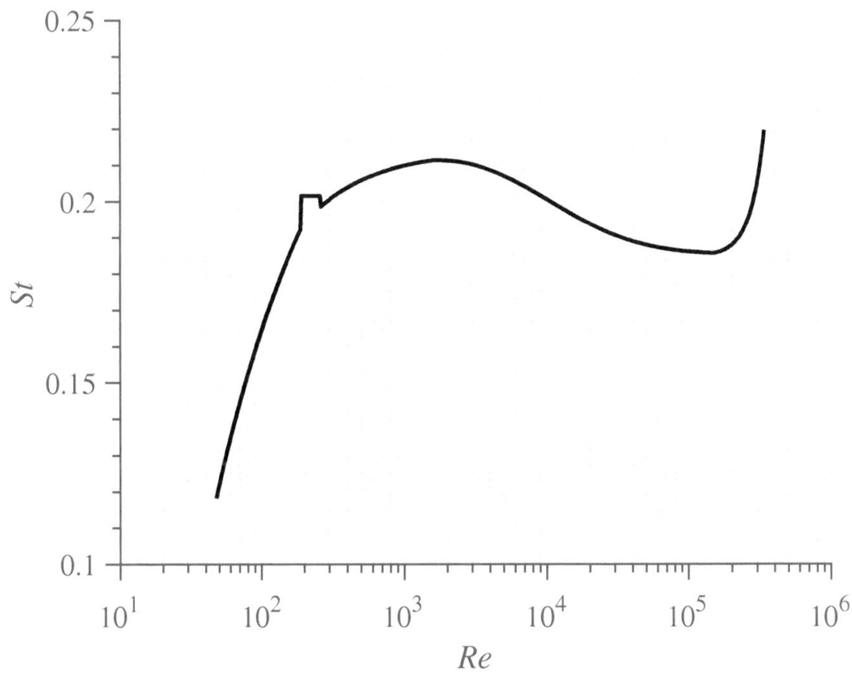

Fig. 7.2 Empirical model of shedding Strouhal number as a function of Reynolds number for a cylindrical cylinder (Reynolds number based on cylinder diameter) [11]

lift. As each shed vortex has opposite circulation, the lift on the cylinder will also change sign in a regular cycle.

Second, there is unsteady drag produced by the changing base pressure of the cylinder. Each shed vortex will have a low-pressure core that will cause an increase in suction at the base of the cylinder that is relieved as it moves away, before re-intensifying as the next vortex is formed. Unlike the lift, this effect does not depend on the circulation of the shed vortices. Instead, the oscillation in base pressure (and thus drag) depends only on the rate of production of shed vortices, which is twice the shedding frequency. Hence, drag oscillates at twice the shedding frequency.

Table 7.2 Root-mean-square sectional lift coefficient empirical functions [11]

$Re =$	$C_{L'} \simeq$	
$47 - 190$	$(\epsilon/30 + \epsilon^2/90)^{1/2}$	$\epsilon = (Re - 47)/47$
$165 - 230$	$0.43 \times (Re/230)$	
$230 - 260$	$0.78 \times (Re/260) - 0.26$	
$260 - 1.6 \times 10^3$	$0.045 + 1.05 \times (1 - Re/1.6 \times 10^3)^{4.5}$	
$1.6 \times 10^3 - 5.4 \times 10^3$	$0.045 + 3.0 \times x^{4.6}$	$x = \log(Re/1.6 \times 10^3)$
$5.4 \times 10^3 - 2.2 \times 10^5$	$0.52 - 0.06 \times x^{-2.6}$	$x = \log(Re/1.6 \times 10^3)$
$2.2 \times 10^5 - 3.4 \times 10^5$	$0.09 + 0.43 \times \exp[-10^5 \times (Re/10^6)^{10}]$	

Finally, there is often a random lift and drag component to the forces on a cylinder if the Reynolds number is high enough for significant turbulence to be present in the wake.

Norberg [11] has developed empirical functions that describe the unsteady root-mean-square sectional lift coefficient of cylinders as a function of Reynolds number. Lift is almost always the dominant force on the cylinder, hence it the most important for analysis of noise and vibration. The sectional lift coefficient is the lift measured on a two-dimensional section of the cylinder. The root-mean-square sectional lift coefficient is defined as,

$$C_{L'} = \frac{l_{rms}}{q_\infty d} \tag{7.85}$$

where l_{rms} is the root-mean-square lift-force-per-unit-span and q_∞ is the freestream dynamic pressure. Equation 7.88 provides a general definition of cylinder force-per-unit-span.

Table 7.2 presents Norberg's empirical functions for sectional root-mean-square lift coefficient. Figure 7.3 presents the empirical functions as a function of Reynolds number.

7.6.1.3 Spanwise Correlation Length

The sectional lift coefficient is expected to have the same magnitude at each section along the span; however, its phase will vary randomly because of the instabilities and turbulence that forms in the wake for Reynolds numbers beyond the laminar-shedding regime. Over a very small length of span, we can expect that the sectional lift forces would be in phase (or correlated) but as the span increases in length, the sectional lift forces will decorrelate and the overall root-mean-square lift will reduce.

The correlation of the forces along the span is measured using a correlation coefficient. If sectional lift (L') can be measured, the normalised correlation coefficient at zero time delay is,

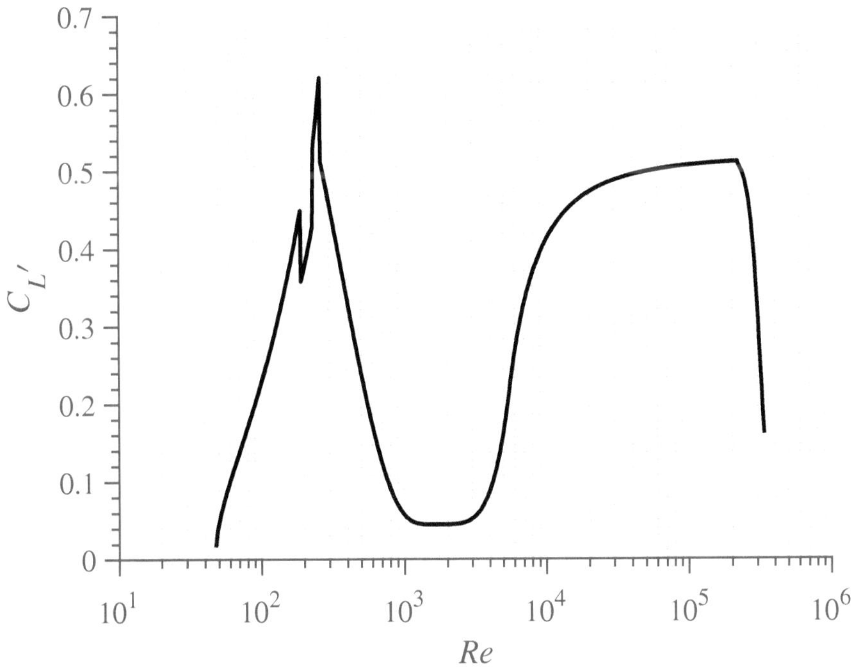

Fig. 7.3 Empirical model of root-mean-square sectional lift coefficient as a function of Reynolds number for a cylindrical cylinder (Reynolds number based on cylinder diameter) [11]

$$
\begin{aligned}
R_{LL}(z_2 - z_1) &= \frac{1}{\sqrt{\overline{L'^2(z_1)L'^2(z_2)}}}\overline{L'(z_1, t)L'(z_2, t)} \\
&= \frac{1}{\sqrt{\overline{L'^2(z_1)L'^2(z_2)}}}\frac{1}{T}\int_0^T L'(z_1, t)L'(z_2, t)\,\mathrm{d}t
\end{aligned} \tag{7.86}
$$

where $R_{LL}(0) = 1$.

While Eq. 7.86 uses sectional lift coefficient, this is sometimes hard to measure or estimate. It is common to calculate the spanwise correlation coefficient using surface pressure measurements or even using wake velocity measurements. A useful study would be to compare these methods and assess their accuracy.

Integrating the normalised correlation coefficient along the span yields the spanwise correlation length Λ_3,

$$
\Lambda_3 = \frac{1}{2}\int_{-\infty}^{\infty} R_{LL}(\eta)\,\mathrm{d}\eta \tag{7.87}
$$

where $\eta = z_2 - z_1$.

Table 7.3 Normalised spanwise correlation length empirical functions [11]

$Re =$	$\Lambda_3/d \simeq$	
$47 - 190$	∞	
$165 - 285$	$11.5 \times [1 - \text{erf}(\mathcal{N})] + 7.0$	$\mathcal{N} = 22 \times (Re/250 - 1)$
$285 - 1.5 \times 10^3$	$7.6 \times (Re/1.5 \times 10^3)^{-0.82}$	
$1.5 \times 10^3 - 1.72 \times 10^3$	7.6	
$1.72 \times 10^3 - 5.1 \times 10^3$	$7.6 \times (Re/1.72 \times 10^3)^{0.60}$	
$5.1 \times 10^3 - 8.0 \times 10^3$	$14.6 \times (Re/5.1 \times 10^3)^{-2.3}$	
$8.0 \times 10^3 - 2.4 \times 10^5$	$2.6 \times (Re/2.4 \times 10^5)^{-0.20}$	
$2.4 \times 10^5 - 3.0 \times 10^5$	$1.4 \times (Re/3.0 \times 10^5)^{-2.7}$	

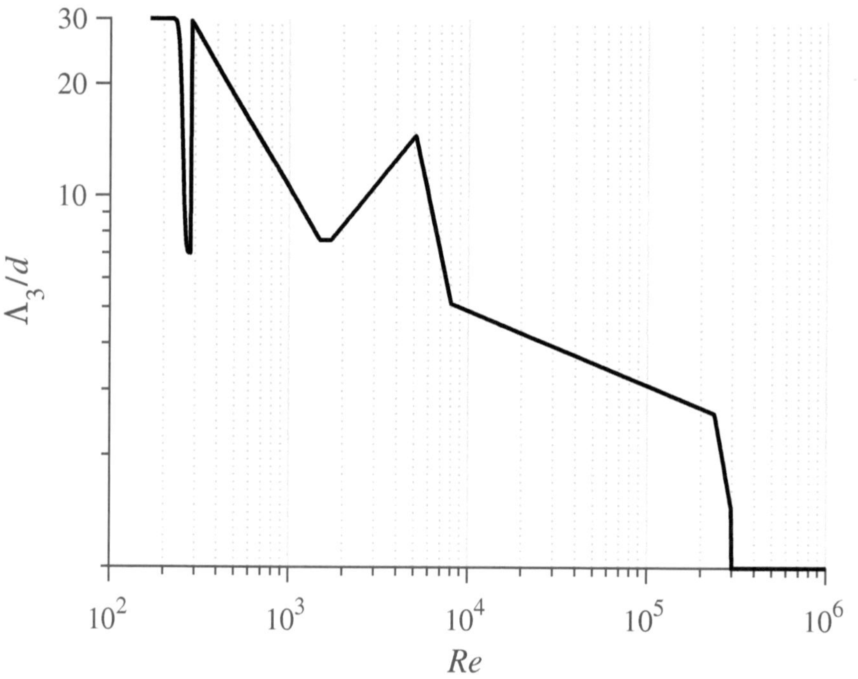

Fig. 7.4 Empirical model of spanwise correlation length scale (normalised by diameter d) as a function of Reynolds number for a cylindrical cylinder (Reynolds number based on cylinder diameter) [11]

As will be shown, this is an important parameter for estimating noise from cylinders in cross flow. Experimental and numerical data have been combined by Norberg [11] to produce empirical functions for the spanwise correlation length scale. Table 7.3 presents these empirical functions and Figure 7.4 plots the empirical model for spanwise correlation length as a function of Reynolds number.

7.6.2 Acoustic Model for a Cylinder in Cross Flow

Let us assume that a cylinder is placed in cross flow. The Reynolds number is large so that there is turbulent vortex shedding that occurs with random phase over the span. Introducing the *force-per-unit-span* f_i,

$$f_i(z,t) = \frac{d}{2}\int_0^{2\pi} P_i(\theta, z, t)\,\mathrm{d}\theta \tag{7.88}$$

where a switch to polar coordinates on the cylinder has been made so that θ is the angle around the circumference of the cylinder and z is the spanwise position along the span of the cylinder ($z = 0$ is the mid-point of the span). Further, d is the diameter of the cylinder. Using Curle's theory and assuming that the cylinder is acoustically compact (Eq. 7.81),

$$p'(\mathbf{x}, t) = \frac{1}{4\pi c_0}\frac{x_i}{r^2}\int_{-L/2}^{L/2}\frac{\partial}{\partial t}[f_i(z,\tau)]_{\tau=\tau^*}\,\mathrm{d}z \tag{7.89}$$

where L is the span of the cylinder.

A useful model for f_i is,

$$f_i(z,t) = |\mathfrak{F}(\omega)|e^{-i(\omega t-\phi(z))} \tag{7.90}$$

where $|\mathfrak{F}(\omega)|$ is a frequency-dependent function of the force-per-unit-span and $\phi(z)$ is the stochastic variation of phase of this force along the span [1, 3]. Taking the Fourier transform of the above creates,

$$\hat{F}(\omega) = |F(\omega)|e^{i\phi(z)} \tag{7.91}$$

where the units of $|F(\omega)|$ are $\frac{\mathrm{N\cdot s}}{\mathrm{m\cdot rad}}$ assuming that T is the sample length used to collect f_i.

Substituting the Eq. 7.90 into Eq. 7.89,

$$p'(\mathbf{x}, t) = \frac{\omega}{4\pi c_0}\frac{x_i}{r^2}\int_{-L/2}^{L/2}\left[|F(\omega)|\,e^{-i(\omega t-\phi(z))}\right]_{\tau=\tau^*}\,\mathrm{d}z \tag{7.92}$$

Taking the Fourier transform,

$$\hat{P}(\mathbf{x}, \omega) = \frac{\omega}{4\pi c_0}\frac{x_i}{r^2}|F(\omega)|\int_{-L/2}^{L/2} e^{i\phi(z)}\,\mathrm{d}z \tag{7.93}$$

and converting to the power spectral density,

$$\Phi_{pp}(\mathbf{x},\omega) = \frac{\pi}{T}E[\hat{P}\hat{P}^*] = \frac{\omega^2}{16\pi^2 c_0^2}\frac{x_i x_i}{r^4}|\mathcal{F}(\omega)|^2 \int_{-L/2}^{L/2}\int_{-L/2}^{L/2} e^{i\phi(z_1-z_2)}\,\mathrm{d}z_1\mathrm{d}z_2 \tag{7.94}$$

where the units of $|\mathcal{F}(\omega)|$ are $\frac{\mathrm{N}\cdot\sqrt{\mathrm{s}}}{\mathrm{m}}$. The integral for phase can be simplified using [3],

$$\int_{-L/2}^{L/2}\int_{-L/2}^{L/2} e^{i\phi(z_1-z_2)}\,\mathrm{d}z_1\mathrm{d}z_2 \approx L\int_{-L/2}^{L/2}\rho(\eta)\,\mathrm{d}\eta \tag{7.95}$$

where $\eta = z_1 - z_2$ and $\rho(\eta)$ is a normalised correlation function. It is reasonable to assume that η is exponential (or Laplacian [3]),

$$\rho(\eta) = e^{-\frac{|\eta|}{\Lambda_3}} \tag{7.96}$$

where Λ_3 is the spanwise correlation length scale. When this is the case,

$$\int_{-L/2}^{L/2}\rho(\eta)\,\mathrm{d}\eta = 2\Lambda_3 \tag{7.97}$$

and the power spectral density for acoustic pressure becomes,

$$\Phi_{pp}(\mathbf{x},\omega) = \frac{\omega^2 L\Lambda_3}{8\pi^2 c_0^2}\frac{x_i x_i}{r^4}|\mathcal{F}(\omega)|^2 \tag{7.98}$$

and the acoustic mean-square-pressure in a small frequency band $\delta\omega$ centred on ω is,

$$\overline{p'^2}(\mathbf{x},\omega) = \frac{\omega^2 L\Lambda_3}{8\pi^2 c_0^2}\frac{x_i x_i}{r^4}|\tilde{F}(\omega)|^2 \tag{7.99}$$

where $\tilde{F}(\omega)$ is the force-per-unit-span of the cylinder and it is reasonable to assume $\tilde{F}(\omega) \approx \mathfrak{F}(\omega)$.

Equation 7.99 now links the flow properties of the cylinder with the acoustic pressure. Namely, we can see the dependence on correlation length scale Λ_3 and the sectional fluctuating force spectrum.

7.6.3 Cylinder Flow Noise Analysis

In this section, empirical models of the various cylinder unsteady flow parameters (see Tables 7.1, 7.2 and 7.3) are used with Equation 7.99 to understand how noise emission varies with Reynolds number. The acoustic pressure is non-dimensionalised,

$$\frac{\overline{p'^2}(\mathbf{x},\omega)}{q_\infty^2} = \frac{1}{2}St^2\left(\frac{L}{d}\right)\left(\frac{\Lambda_3}{d}\right)M_\infty^2 D_i\left(\frac{d^2}{r^2}\right)\tilde{C}_{F,i}(\omega)^2 \tag{7.100}$$

where D_i is a directivity function,

$$D_i = \frac{x_i x_i}{r^2} \tag{7.101}$$

and the Strouhal number is defined,

$$St = \frac{\omega d}{2\pi U_\infty} \tag{7.102}$$

and the force coefficient is,

$$\tilde{C_F}(\omega) = \frac{|\tilde{F}(\omega)|}{q_\infty d} \tag{7.103}$$

Equation 7.100 is rearranged further to obtain an acoustic pressure coefficient, K, for sound produced at the vortex shedding frequency,

$$\mathrm{K} = 10\log_{10}\left[\frac{2\overline{p_a'^2}r^2}{q_\infty^2 M_\infty^2 L d}\right] = 10\log_{10}\left[St^2\left(\frac{\Lambda_3}{d}\right)C_{L'}^2\right] \tag{7.104}$$

Here, we assume the measurement position is directly above the cylinder so that $D_i = 1$ and $\tilde{C_F}(\omega) = C_{L'}$.

Figure 7.5 presents the acoustic pressure coefficient as defied by Eq. 7.104. The shedding frequency, spanwise correlation length and root-mean-square lift coefficient combine to affect the sound level in the far field. At low Reynolds number ($Re \lesssim 290$), the acoustic pressure is mostly high due to the high spanwise correlation length. As Reynolds number increases, the acoustic pressure coefficient decreases rapidly until $Re \approx 1.5 \times 10^3$. This is primarily due to the rapidly reducing unsteady lift, but also because of the reducing spanwise correlation length. As Reynolds number increases, so does the acoustic pressure coefficient because unsteady lift and the spanwise correlation length both increase. Over the range $8000 \lesssim Re \lesssim 2.4 \times 10^5$, the acoustic pressure coefficient is approximately constant because the unsteady lift increase balances the spanwise correlation length decrease. The acoustic pressure decreases for $Re \gtrsim 2.4 \times 10^5$ as lift and spanwise correlation length both drop suddenly.

7.7 The Ffowcs Williams and Hawkings Equation

Up until this point, our acoustic sources have been stationary. Ffowcs Williams and Hawkings [7] developed a general equation for acoustic sources generated by moving sources.

Consider a moving surface of area $S_0(t)$ and volume $V_0(t)$ with velocity $\mathbf{V}$ surrounded by an acoustic medium. Turbulent flow and objects creating sound (by

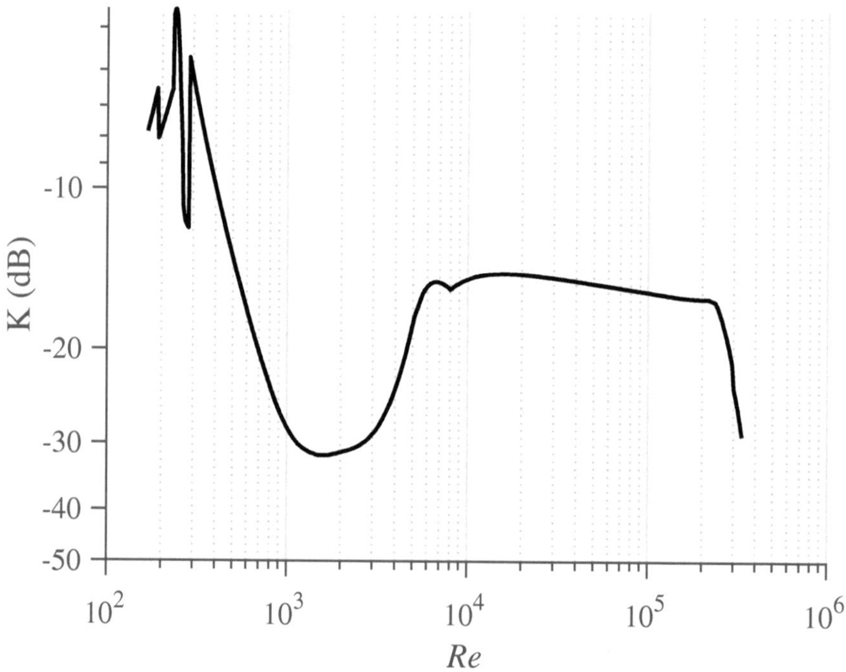

Fig. 7.5 Empirical model of the acoustic pressure coefficient (Eq. 7.104) at the vortex shedding frequency as a function of Reynolds number for a cylindrical cylinder (Reynolds number based on cylinder diameter)

various means) may exist within $V_0(t)$. Ffowcs Williams and Hawkings [7] developed a method to calculate the sound radiated from $V_0(t)$ by considering the surface only. They do this by defining the following smooth generalised (arbitrary) function $f(\mathbf{x}, t)$ so that

$$f(\mathbf{x}, t) = \begin{cases} < 0 & \in V_0(t) \\ = 0 & \in S_0(t) \\ > 0 & \notin V_0(t) \end{cases} \tag{7.105}$$

The Heaviside function $H(f)$ is used to filter the solution so that it is zero within the volume and unaffected outside of it. Mathematically it is defined as,

$$H(f) = \begin{cases} 0 & f < 0 \\ 1 & f > 0 \end{cases} \tag{7.106}$$

The Heaviside function is multiplied by the flow variables p', ρ', $\mathbf{v}$, etc. The continuity and momentum equations are evaluated in terms of these new flow variables to eventually form a new non-homogeneous wave equation. As will be shown, the source terms of this wave equation are surface sources which arise naturally from the mathematical derivation. They occur to ensure that the continuity and momentum equations are conserved. Physically, these represent volume (monopole) and momentum (pressure or force dipole) sources in the flow.

First, the normal $\mathbf{n}$ of the surface $S_0(t)$ must be found. It is defined as,

$$\mathbf{n} = \left[\frac{\nabla f}{\|\nabla f\|} \right]_{f=0} \tag{7.107}$$

On the surface, $\mathbf{x} = \mathbf{x_s}$, $f(\mathbf{x} = \mathbf{x_s}, t) = 0$ and we can write, [5, 8, 14],

$$\frac{\partial f}{\partial t} = -\mathbf{V} \cdot \nabla f = -(\mathbf{V} \cdot \mathbf{n})\|\nabla f\| \tag{7.108}$$

Further, the spatial derivative of H can be written as a Dirac delta function [8],

$$\delta(x) = \frac{\partial H(x)}{\partial x} \tag{7.109}$$

Hence,

$$\nabla H(f) = \frac{\partial H(f)}{\partial f} \nabla f = \delta(f)\|\nabla f\|\mathbf{n} \tag{7.110}$$

We can now combine the Heaviside function $H(f) = H$ with continuity (Eq. 7.55) and conservation of momentum (Eq. 7.56) equations, effectively filtering them so that the flow variables are only non-zero outside of $S_0(t)$,

$$H \left[\frac{\partial \rho}{\partial t} + \frac{\partial (\rho v_i)}{\partial x_i} \right] = 0 \tag{7.111}$$

$$H \left[\frac{\partial (\rho v_i)}{\partial t} + \frac{\partial (\rho v_i v_j + p_{ij})}{\partial x_j} \right] = 0 \tag{7.112}$$

Combining Eqs. 7.111 and 7.112 with Eqs. 7.108 and 7.110 and rearranging we obtain an updated continuity equation,

$$\frac{\partial (\rho' H)}{\partial t} + \frac{\partial (\rho v_i H)}{\partial x_i} = (\rho v_j - \rho' V_j) n_j \delta(f)\|\nabla f\| \tag{7.113}$$

and momentum equation,

$$\frac{\partial (\rho v_i H)}{\partial t} + \frac{\partial (\rho v_i v_j H)}{\partial x_j} = \left[\rho v_i (v_j - V_j) + p_{ij}\right] n_j \delta(f)\|\nabla f\| \tag{7.114}$$

Using the same procedure used in Sect. 7.2.1, we can obtain the Ffowcs Williams and Hawkings [7] wave equation using the new flow variable $\rho' H$,

$$\begin{aligned}\frac{\partial^2(\rho' H)}{\partial t^2} - c_0^2 \frac{\partial^2(\rho' H)}{\partial x_i^2} = {} & \frac{\partial^2(T_{ij} H)}{\partial x_i \partial x_j} \\ & - \frac{\partial}{\partial x_i}\left\{\left[\rho v_i \left(v_j - V_j\right) + p_{ij}\right] n_j \delta(f) \|\nabla f\|\right\} \\ & + \frac{\partial}{\partial t}\left[\left(\rho v_j - \rho' V_j\right) n_j \delta(f) \|\nabla f\|\right]\end{aligned} \tag{7.115}$$

This is a non-homogeneous wave equation with right-hand-side source terms that take into account moving surfaces. It can also be written in terms of acoustic pressure $p' H$,

$$\begin{aligned}\frac{1}{c_0^2}\frac{\partial^2(p' H)}{\partial t^2} - \frac{\partial^2(p' H)}{\partial x_i^2} = {} & \frac{\partial^2(T_{ij} H)}{\partial x_i \partial x_j} \\ & - \frac{\partial}{\partial x_i}\left\{\left[\rho v_i \left(v_j - V_j\right) + p_{ij}\right] n_j \delta(f) \|\nabla f\|\right\} \\ & + \frac{\partial}{\partial t}\left[\left(\rho v_j - \rho' V_j\right) n_j \delta(f) \|\nabla f\|\right]\end{aligned} \tag{7.116}$$

A Green's function (Eq. 7.2) can be used to obtain the solution to the above [7, 8],

$$\begin{aligned}p'(\mathbf{x}, t) = {} & \frac{\partial^2}{\partial x_i \partial x_j} \int_{V_E} \left[\frac{T_{ij}}{4\pi r |1 - M_r|}\right]_{\tau=\tau^*} \mathrm{d}V(\boldsymbol{\eta}) \\ & - \frac{\partial}{\partial x_i} \int_{S_0} \left[\frac{\left[\rho v_i \left(v_j - V_j\right) + p_{ij}\right] n_j}{4\pi r |1 - M_r|}\right]_{\tau=\tau^*} \mathrm{d}S(\boldsymbol{\eta}) \\ & + \frac{\partial}{\partial t} \int_{S_0} \left[\frac{\left(\rho v_j - \rho' V_j\right) n_j}{4\pi r |1 - M_r|}\right]_{\tau=\tau^*} \mathrm{d}S(\boldsymbol{\eta})\end{aligned} \tag{7.117}$$

where V_E is the volume external to V_0. The coordinates of the surface, $\boldsymbol{\eta}$, are defined in a Lagrangian (moving) reference frame so that the sources appear at rest. If $\mathbf{y}$ are the source-fixed coordinates, then $\boldsymbol{\eta}$ is defined by,

$$\mathbf{y} = \boldsymbol{\eta} + \int^{\tau} \mathbf{V}(\tau')\, \mathrm{d}\tau' \tag{7.118}$$

where τ is the emission time. Now r is a function of τ,

$$r(\tau) = |\mathbf{x} - \mathbf{y}| = \left|\mathbf{x} - \boldsymbol{\eta} - \int^{\tau} \mathbf{V}(\tau')\, \mathrm{d}\tau'\right| \tag{7.119}$$

The relative Mach number, M_r, is the Mach number of the component of $\mathbf{V}$ aligned with $(\mathbf{x} - \mathbf{y})$,

$$M_r = \frac{\mathbf{V} \cdot (\mathbf{x} - \mathbf{y})}{r c_0} \tag{7.120}$$

Note that there is no need to use the Heaviside function H in Eq. 7.117 now as the integrals are on the surface or in the volume external to the surface.

7.7.1 Solid Surfaces

The Ffowcs-Williams Hawkings (often abbreviated as FWH) equation is used to predict sound generated by solid moving objects. For solid objects (impermeable), the following condition must be satisfied,

$$\begin{aligned} \mathbf{V} \cdot \mathbf{n} &= \mathbf{v} \cdot \mathbf{n} \text{ or} \\ V_i n_i &= v_i n_i \end{aligned} \tag{7.121}$$

This is often the case for propeller and helicopter rotor blades. If we also consider the additional situation where the sources are compact and in the far-field, the compact multipole solutions described in Sect. 7.1.3 and onward can be used. When this is done, we obtain,

$$\begin{aligned} p'(\mathbf{x}, t) \approx &\frac{x_i x_j}{|r|^3} \frac{1}{c^2} \frac{\partial^2}{\partial t^2} \int_{V_E} \left[\frac{T_{ij}}{4\pi |1 - M_r|} \right]_{\tau=\tau^*} \mathrm{d}V(\eta) \\ &+ \frac{x_i}{|r|^2} \int_{S_0} \left[\frac{p_{ij} n_j}{4\pi |1 - M_r|} \right]_{\tau=\tau^*} \mathrm{d}S(\eta) \\ &+ \frac{1}{r} \frac{\partial}{\partial t} \int_{S_0} \left[\frac{\rho_\infty V_j n_j}{4\pi |1 - M_r|} \right]_{\tau=\tau^*} \mathrm{d}S(\eta) \end{aligned} \tag{7.122}$$

Thus we see there are three main sources of noise associated with a rotor blade and these are associated with the three separate terms on the right hand side of Eq. 7.122. The first is a quadrupole source term associated with turbulence in the volume outside of the blade. This is normally the flow generated by the movement of the blade and turbulence created by the boundary layer. The first term may also be associated with shock waves, such as those produced over the surfaces of transonic blade tips. The second term is a dipole source term associated with the forces acting on the rotating blades. If the flow were inviscid, this may be called "Gutin" noise [1]. The final term is a monopole source that is created by the volume displacement of the surrounding medium by the moving blade. This is called thickness noise.

7.8 Summary

This chapter has covered the fundamentals of how unsteady flow creates sound. After a statement of some background theory and definitions, Lighthill's equation is presented, along with its solution. This is followed with a general discussion of noise form unsteady mass injection and force. After this, the effect of solid surfaces is illustrated by describing Curle's theory for stationary surfaces followed by the Ffowcs-Williams and Hawkings equation for moving surfaces. An analysis of cylinder flow and noise is provided.

References

1. Blake WK (2017) Mechanics of flow-induced sound and vibration, Volume 1: General concepts and elementary sources, 2nd edn. Academic, London
2. Blake WK (2017) Mechanics of flow-induced sound and vibration, Volume 2: Complex flow-structure interactions. Academic
3. Casalino D, Jacob M (2003) Prediction of aerodynamic sound from circular rods via spanwise statistical modelling. J Sound Vib 262(4):815–844. https://doi.org/10.1016/S0022-460X(02)01136-7, https://linkinghub.elsevier.com/retriev
4. Curle N (1955) The influence of solid boundaries upon aerodynamic sound. Proc. Roy. Soc. Lond. Ser. A Math. Phys. Sci. 231(1187):505–514
5. Delfs J (2016) Grundlagen der Aeroakustik (Basics of Aeroacoustics). Technische Universität Braunschweig, Braunschweig
6. Dowling AP, Mahmoudi Y (2015) Combustion noise. Proc. Combust. Inst. 35(1):65–100
7. Ffowcs-Williams J, Hawkings D (1969) Sound generation by turbulence and surfaces in arbitrary motion. Philos. Trans. Roy. Soc. Lond. Ser. A Math. Phys. Sci. 264(1151):321–342. https://doi.org/10.1098/rspa.1952.0060
8. Glegg S, Devenport W (2017) Aeroacoustics of low Mach number flows: fundamentals, analysis, and measurement. Academic, New York
9. Howe MS (1998) Acoustics of fluid-structure interactions. Cambridge University Press
10. Lighthill MJ (1952) On sound generated aerodynamically I. General theory. Proc. Roy. Soc. Lond. Ser. A Math. Phys. Sci. 211(1107):564–587, https://doi.org/10.1098/rspa.1952.0060
11. Norberg C (2003) Fluctuating lift on a circular cylinder: review and new measurements. J Fluids Struct 17(1):57–96. https://doi.org/10.1016/S0889-9746(02)00099-3
12. Stratton J (1941) Electromagnetic theory mcgraw-hill. New York pp 31–434
13. Strouhal V (1878) Über eine besondere Art der Tonerregung. Stahel
14. Wagner C, Hüttl T, Sagaut P (2007) Large-eddy simulation for acoustics, vol 20. Cambridge University Press
15. Williamson CHK (1996) Vortex dynamics in the cylinder wake. Ann Rev Fluid Mech 28(1):477–539 , https://doi.org/10.1146/annurev.fl.28.010196.002401

Chapter 8
Airfoil Noise Mechanisms and Control

Abstract A common component responsible for unwanted noise in low Mach number flows is the airfoil. Airfoil noise, produced when unsteady fluid flow interacts with an airfoil surface, is important in many applications ranging from wind turbines and cooling fans to aircraft and submarines. This chapter presents an introduction to the major mechanisms of airfoil noise production. We begin with a review of leading edge noise that is produced by the interaction of upstream turbulent flow with the leading edge of an airfoil. Following this, we will discuss two distinct types of trailing edge noise. The first is broadband noise produced when boundary layer turbulence that forms on the surface of the airfoil interacts with the sharp trailing edge. Turbulent fluctuations are scattered by the edge and this increases their acoustic radiation efficiency. The second source of trailing edge noise is tonal in nature and can be generated by either vortex shedding or an aeroacoustic feedback loop. Our intention is for the reader to gain physical understanding of the most important airfoil noise generation mechanisms. As part of our discussion, brief reviews of several analytical leading and trailing edge noise models are presented to provide insight into how noise is created and can be predicted. The chapter concludes with an overview of several different passive noise control devices that can be used to control leading and trailing edge noise. These devices, including serrations, porosity, brushes and finlets, are inspired by the phenomenon of silent owl flight and seek to replicate unique features of the owl wing. We will review current information on the effectiveness of passive noise control devices along with our current understanding of their underlying noise reduction mechanisms.

8.1 Leading Edge Noise

Broadband noise is created when an upstream turbulent flow impinges on an airfoil leading edge, as illustrated in Fig. 8.1. This type of noise, known as leading edge or airfoil-turbulence interaction noise, is important for a variety of technologies including wind turbines, aircraft engines, helicopter rotors and submarine propellers. Wind turbine blades encounter atmospheric boundary layer turbulence and the wakes shed by upstream turbines resulting in the production of leading edge noise. In the

C. Doolan and D. Moreau, *Flow Noise*,
https://doi.org/10.1007/978-981-19-2484-2_8

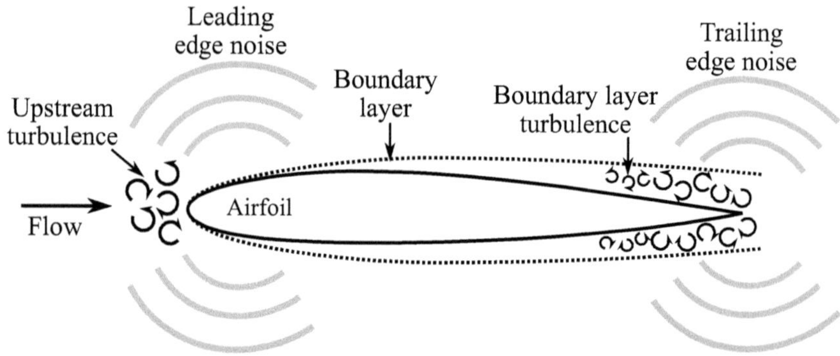

Fig. 8.1 The airfoil flow and noise environment showing leading edge and broadband trailing edge noise production

case of a submarine propeller, the turbulence impinging on the leading edge is created by the vehicle boundary layer and flows over inlet guide vanes. The stators in an aeroengine encounter wake turbulence from upstream fan blades and the boundary layer that forms on the outer wall.

When turbulent eddies (created by atmospheric shear, a wall boundary layer or by other components upstream of the airfoil) encounter the airfoil leading edge, the random velocity fluctuations in the flow induce an unsteady pressure over the surface of the airfoil. The unsteady pressure distribution creates unsteady lift which is a source of sound according to the theory of Curle [37] (see Sect. 7.5). Amiet [1] has derived a model that describes this noise generation process. In the model, leading edge noise is deduced from a statistical description of the incident turbulent flow field using a transfer function between the turbulent flow field and the resulting lift. The unsteady lift is then converted to far-field sound using the theories of Kirchoff [81] and Curle [37].

Consider an airfoil with span $2d$ and chord $2c$ placed in turbulent flow. The origin of the co-ordinate system is located at the airfoil midchord and midspan. The x co-ordinate corresponds to the chordwise direction, the y co-ordinate corresponds to the spanwise direction and the z co-ordinate corresponds to the vertical (airfoil-normal) direction. Following Amiet's formulation [1] (see also Paterson and Amiet [107]; Moreau and Roger [96]), the far-field acoustic Power Spectral Density (PSD) in the midspan ($y = 0$) plane due to turbulence interaction with the leading edge is given by

$$S_{pp}(\mathbf{x}, \omega) = \left(\frac{\omega z \rho_0 c M}{\sigma^2}\right)^2 d|L|^2 \phi_{ww}(\omega)\Lambda_y(\omega) \tag{8.1}$$

where $\mathbf{x} = (x, 0, z)$ is the observer location, $\sigma = \sqrt{x^2 + \beta^2(y^2 + z^2)}$ is the far-field corrected observer location, M is the Mach number and $\beta = \sqrt{1 - M^2}$ is the compressibility term. The term $\phi_{ww}(\omega)$ is the spectrum of the vertical velocity fluctuations, $\Lambda_y(\omega)$ is the spanwise correlation length scale of the vertical velocity fluctuations and L is the airfoil response function that relates the fluctuating lift to noise. Note that to calculate the physically realisable one-sided PSD defined for positive frequencies only, we calculate $G_{pp} = 2S_{pp}$.

Amiet's theory [1] was formulated based on three main assumptions. First, the incoming turbulence is assumed to be frozen, termed the "frozen turbulence assumption". This means that the incident turbulent velocity field is unaffected by the presence of the leading edge and it is unmodified when it convects past the leading edge. Second, the airfoil is assumed to be an infinitely thin flat plate at zero angle of attack, termed the "thin plate assumption". Lastly, end effects are ignored, referred to as the "large aspect ratio assumption", such that the airfoil is considered to have infinite span in the calculation of the airfoil response.

If the PSD of the vertical velocity fluctuations and the spanwise correlation length scale of the incident flow field are not already known, the von Kármán or Liepmann models of isotropic turbulence may be used instead (see Sects. 6.7.1 and 6.7.2).

Different expressions for the airfoil response function exist at low and high frequencies [1]. The low frequency solution is valid for $\frac{M\hat{\omega}}{\beta^2} < \frac{\pi}{4}$ where $\hat{\omega}$ is the reduced frequency $\omega c/U_\infty$. According to Paterson and Amiet [107], the low frequency airfoil response function can be approximated as

$$|L(\hat{\omega})| = \frac{1}{\beta} S(\hat{\omega}/\beta^2) \tag{8.2}$$

where S is the Sears function [117]. An accurate approximation to the Sears function which can be used in Eq. 8.2 is given by [107]

$$S(k) \approx \left[\frac{1}{1+2.4k} + 2\pi k\right]^{-1/2} \tag{8.3}$$

At high frequencies when $\frac{M\hat{\omega}}{\beta^2} > \frac{\pi}{4}$, the airfoil response function is divided into two parts given by

$$L_1(\hat{\omega}) = \frac{1-i}{\pi\hat{\omega}\sqrt{M}} \mathbf{E}^* \left(\sqrt{\frac{4\hat{\omega}M}{\pi(1+M)}}\right) \tag{8.4}$$

and

$$\begin{aligned} L_2(\hat{\omega}) = &\frac{\sqrt{1+M}}{iM(\pi\hat{\omega})^{3/2}} \left[\mathbf{E}^*\left(\frac{2}{\beta}\sqrt{2\hat{\omega}M/\pi}\right) - \frac{1-i}{2}\right. \\ &\left. + \left[\frac{1-i}{2} - \sqrt{\frac{2}{1+M}}\mathbf{E}^*\left(\sqrt{\frac{4\hat{\omega}M}{\pi(1-M)}}\right)\right] e^{i(2\hat{\omega}M/(1+M))}\right] \end{aligned} \tag{8.5}$$

where $\mathbf{E}^*(x) = C(x) - iS(x)$ and $C(x)$ and $S(x)$ are Fresnel cosine and sine integrals, respectively. The total airfoil lift response function is then

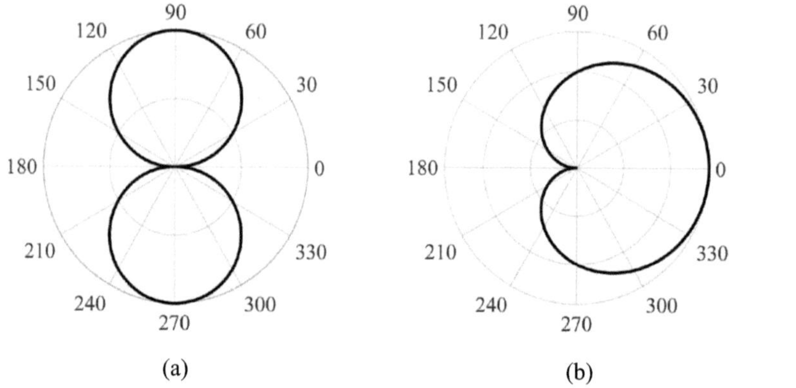

Fig. 8.2 Leading edge noise directivity patterns. **a** Dipole and **b** cardioid radiation pattern. The origin is at the leading edge location and the flow direction is from left to right

$$L = L_1 + L_2 \tag{8.6}$$

At very high frequencies when $\hat{\omega} \to \infty$, an asymptotic expression for the airfoil lift response function can be found as [96]

$$L(\hat{\omega}) = L_1(\hat{\omega}) \to \frac{-i}{\pi \hat{\omega} \sqrt{M}} \tag{8.7}$$

The directivity pattern of leading edge noise depends on the size or wavelength of the incoming turbulent gust. At low frequencies when the size of the turbulent eddy approaching the leading edge is large compared to the airfoil chord, the leading edge noise exhibits a dipole radiation pattern as illustrated in Fig. 8.2a. At high frequencies when the eddy size is much smaller than the airfoil chord, the directivity pattern tends towards cardioid-like shape (see Fig. 8.2b). Amiet's model [1] also predicts the mean velocity dependence of the radiated sound is to the fifth power at low frequencies and the sixth power at high frequencies [106].

While predictions of leading edge noise are often based on Amiet's model [1] in practice (see for example Bertagnolio et al. [13]; Buck et al. [19]), the effects of real airfoil geometry on leading edge noise have not yet been fully addressed. Several authors [40, 51, 90, 95, 111] have examined the influence of airfoil thickness, camber and angle of attack on leading edge noise production. The thin plate assumption has been found to hold in practice for airfoils with thickness below a few percent of the chord and incoming disturbance levels below 10% [111]. Increasing the airfoil thickness attenuates the high frequency content of the radiated sound and Amiet's model may overpredict the noise levels of thick airfoils by more than 10 dB at high frequencies and low flow speeds [40, 111]. Angle of attack and camber have been shown to only have minor influence on leading edge noise production under the assumption of homogenous and isotropic turbulence [40]. It has been speculated however, that anisotropic turbulent inflow could induce significant angle of attack and airfoil geometry effects [40, 111] but further work is needed to understand this.

Mish and Devenport [91, 92] related the sectional lift of a NACA0015 airfoil to unsteady surface pressure to examine angle of attack effects on the airfoil response. The sectional lift (which differs from the total lift spectrum by a spanwise correlation length scale dependent on frequency) and pressure spectral levels displayed a reduction of up to 5 dB with increasing angle of attack for reduced frequencies $\hat{\omega}_c < 5$ (based on the convective velocity, U_c, of eddies approaching the airfoil). For $\hat{\omega}_c > 5$, a significant increase in the pressure spectral levels was observed. The reduction in lift at low frequencies was attributed to distortion of the inflow caused by the increased rate of strain imposed by the leading edge on the flow. It was shown that angle of attack effects arising from inflow distortion are significant only when the relative scale of the turbulent eddies to the airfoil chord is small ($\lambda/c = 13\%$ for the experiments of [91, 92]). The high frequency increase in spectral levels was attributed to a mean-loading effect where thickening of the boundary layer on the suction side of the airfoil at higher angle of attack results in the production of large-scale eddies and high amplitude pressure fluctuations. Using a panel method calculation [51], Devenport et al. [40] also showed that angle of attack has a significant effect on the airfoil lift response function due to inflow distortion at the leading edge. However, when the response function is averaged with the isotropic turbulence spectrum, angle of attack effects on leading edge noise production become small.

To account for thickness effects, several authors [95, 115] have used Rapid Distortion Theory (RDT) [71] to model distortion during turbulence-interaction with a leading edge of finite radius of curvature. Santana et al. [115] used RDT to modify the von Kármán turbulence model and comparison was made with near-wall turbulence measurements taken close to the leading edge of a NACA0012 airfoil. Turbulence distortion was found to occur in a small upstream region (the order of magnitude of the leading edge radius of curvature) and RDT showed some promise for modelling this effect. Improved leading edge noise prediction was obtained when the RDT modified turbulence spectrum was used with Amiet's [1] theory instead of the original turbulence model. However, further work remains to determine the optimal turbulent inflow measurement position for accurate leading edge noise prediction.

8.2 Trailing Edge Noise

The production of sound by airfoils operating at low Mach number is often dominated by noise generated at the trailing edge. In aeronautical applications, this noise source is produced by turbo-fan engines and components of the airframe. In naval applications, the noise generated by the trailing edges of marine propellers and hydrofoils is of serious concern. Trailing edge noise is also created by wind turbines, axial and centrifugal fans in rotating machines, and helicopter and UAV rotors.

Trailing edge noise can be categorised as tonal or broadband. Tonal noise can be produced by a feedback loop mechanism in a laminar-transitional flow regime or when blunt trailing edge vortex shedding occurs. Broadband noise is generated at sufficiently high Reynolds number (typically $Re \geq 5 \times 10^5$, based on chord), when turbulent boundary layers form on the surface of the airfoil and interact with a sharp trailing edge. The tonal and broadband mechanisms of trailing edge noise are discussed as follows.

8.2.1 Laminar-Transitional Boundary Layer Tonal Noise

At low-to-moderate Reynolds number (typically $Re \leq 5 \times 10^5$, based on chord), a unique type of tonal noise is produced by airfoils with a laminar or laminar-transitional boundary layer and a sharp trailing edge. An example of the airfoil tonal noise spectrum is given in Fig. 8.3. It consists of a broadband "hump" with centre frequency f_s superimposed with a primary or dominant tone referred to as $f_{n,max}$ and multiple equispaced tones denoted f_n [4, 108]. A notable feature of the primary tone is its variation in frequency with free-stream velocity. Using NACA0012 and NACA0018 airfoils, Paterson et al. [108] observed that the primary tone frequency increases according to $U_\infty^{0.8}$ over a finite flow speed range before "jumping" to a higher frequency and following a new 0.8 power law with free-stream velocity. This behaviour is referred to as the "ladder-type" frequency structure where the discontinuous steps constitute individual rungs of the ladder and is illustrated in Fig. 8.4. The overall frequency behavior of the tones typically follows the power law of $U_\infty^{1.5}$ and an empirical relation describing the mean behaviour of the primary tone frequency is [4, 108]

$$f_{n,max} = \frac{0.011 U_\infty^{1.5}}{\sqrt{2c\nu}} \tag{8.8}$$

The empirical relation of Eq. 8.8 can also be used to describe the centre frequency of the broadband noise contribution, f_s, as the broadband contribution on which the tones are superimposed determines the primary tone frequency. Arbey and Bataille [4] also formulated an empirical relationship for f_s that is given by

$$f_s = \frac{St_s U_\infty}{\delta_{TE}^*} \tag{8.9}$$

where $St_s = 0.048 \pm 0.003$ and δ_{TE}^* is the boundary layer displacement thickness at the trailing edge. This equation is in reasonable agreement with Eq. 8.8.

The earliest studies of tonal trailing edge noise concluded that it is produced by vortex shedding from the trailing edge [108]. Subsequent studies however, have attributed tonal noise generation to an aeroacoustic feedback loop between flow instabilities in the boundary layer and acoustic waves. A schematic of the aeroacoustic feedback loop mechanism is shown in Fig. 8.5. The flow instabilities are Tollmein-Schlicting (T-S) waves that occur naturally in a boundary layer that is undergoing transition to turbulence. The T-S waves comprise many wavelengths (or frequencies)

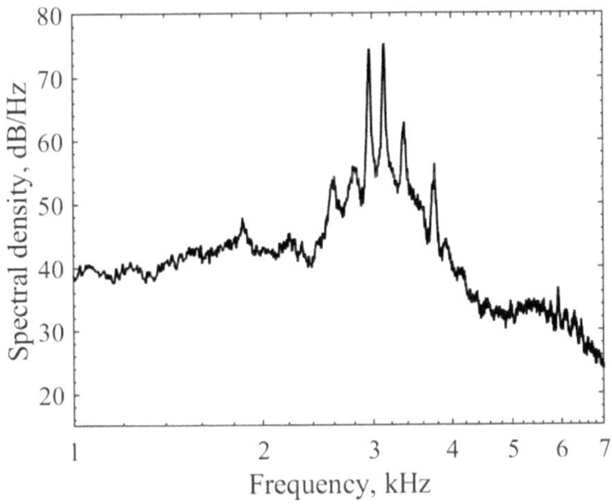

Fig. 8.3 Example laminar-transitional boundary layer tonal noise spectrum for a NACA0012 airfoil at an angle of attack of 5° and $Re = 1.6 \times 10^5$

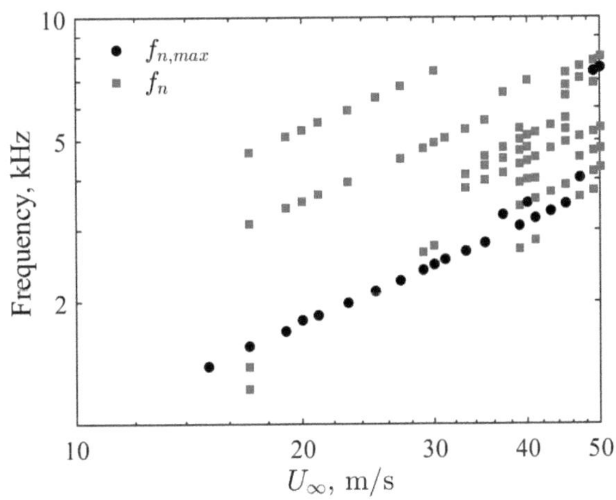

Fig. 8.4 Example of the ladder-type frequency structure of the tonal noise frequencies for a NACA0012 airfoil at zero angle of attack and $Re = 6 \times 10^4 - 2.3 \times 10^5$

and produce a broadband hump in the wall pressure pressure spectrum, the shape of which can be predicted using laminar boundary layer linear stability theory. Tam [119] first proposed that tonal noise is produced by an aeroacoustic feedback loop between the first point of boundary layer instability and a point in the wake which acts as the noise source. This aeroacoustic feedback loop model has since been modified by a number of researchers [4, 5, 39, 77, 80, 109, 116, 124], who have suggested that a feedback loop between instabilities in the boundary layer and acoustic waves generated at the trailing edge is responsible for the tonal noise. The T-S waves are diffracted by the trailing edge producing acoustic waves whereby the trailing edge wall pressure spectrum and the far-field noise spectrum exhibit the same broadband hump with centre frequency, f_s. The acoustic waves travel upstream to reinforce the flow instabilities at their source if both the acoustic signal and the T-S waves are in

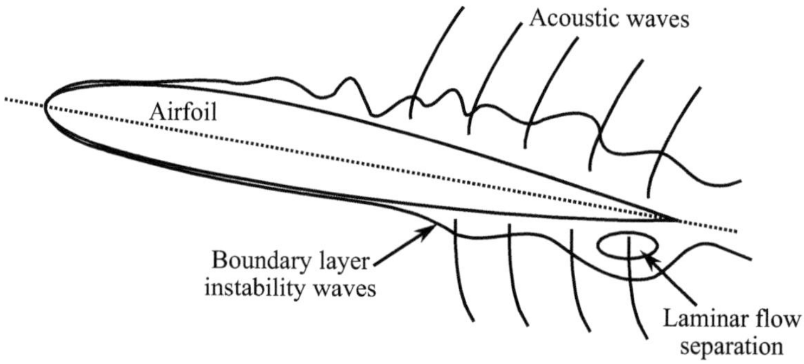

Fig. 8.5 Aeroacoustic feedback loop mechanism responsible for laminar-transitional boundary layer tonal noise

phase at that point. This leads to an amplification of selected T-S waves above that of their original broadband level.

Arbey and Bataille [4] proposed that the tonal noise frequencies, f_n, may be predicted using an aeroacoustic feedback loop model with feedback loop length, L, according to

$$\frac{f_n L}{U_c}\left(1+\frac{U_c}{c_0-U_\infty}\right)=n+\frac{1}{2} \tag{8.10}$$

where U_c is the convection velocity and n is a positive integer value. In their original formulation, Arbey and Bataille [4] used a value of L equal to the distance between the trailing edge and the point of maximum flow velocity over the airfoil surface. They also derived an empirical formula for the frequency spacing Δf between tones f_n based on feedback loop length L given by

$$\Delta f=\frac{K U_\infty^m}{L} \tag{8.11}$$

where $K=0.89 \pm 0.05$ and $m=0.85 \pm 0.01$.

Tonal noise production has also been associated with airfoil flows that exhibit laminar flow separation near the trailing edge. Using linear stability analysis and experimentally measured boundary layer profiles on the airfoil pressure surface, McAlpine et al. [89] and Nash et al. [100] calculated the amplification of T-S waves over the pressure surface. They demonstrated that T-S wave amplification occurs across a small laminar separation bubble located near to the trailing edge and the frequency of the T-S wave that undergoes maximum amplification coincides with the frequency of the tonal noise. McAlpine et al. [89] proposed that an aeroacoustic feedback loop is not required for the generation of tonal noise; it can be explained in terms of the amplification of boundary layer instabilities only. In contrast, feedback loop models between an acoustic source at the trailing edge and points of boundary layer separation where T-S wave amplification occurs have been proposed by several

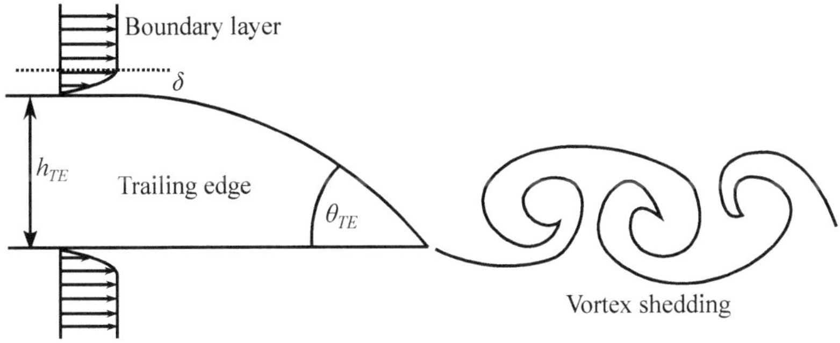

Fig. 8.6 Blunt trailing edge vortex shedding from a bevelled trailing edge

authors [5, 39]. Extending the work of McAlpine et al. [89] and Nash et al. [100], Kingan and Pearse [80] used T-S wave amplification theory to calculate the phase change between the first point of flow instability and the trailing edge. They incorporated this information into the feedback loop model of Arbey and Bataille [4] to create a new predictive tool for tonal noise frequency selection.

Based on the preceding discussion, readers may be aware that the scientific community is yet to reach consensus on the mechanism responsible for airfoil tonal noise production. Some debate still remains about the feedback loop length L (or the point at which the acoustic wave radiated upstream from the trailing edge is positively reinforced) or even whether an aeroacoustic feedback loop is a necessary condition for the production of airfoil tonal noise. Some authors believe that the feedback loop is a facility induced acoustic reflection effect that can be eliminated with appropriate acoustic treatment around the airfoil [100, 120]. As such, the airfoil tonal noise mechanism continues to be an open question in academic research.

8.2.2 Blunt Trailing Edge Tonal Noise

A trailing edge with large bluntness in comparison to the boundary layer thickness generates vortex shedding in the wake and this in turn leads to the production of tonal noise. The term "blunt" refers to an edge of finite thickness and includes such edge geometries as truncated, rounded and asymmetrically bevelled. A bevelled edge, as illustrated in Fig. 8.6, is commonly characterized by the angle enclosed by the surfaces at the trailing edge, θ_{TE}.

Blunt trailing edge vortex shedding can produce a narrowband tone that is orders of magnitude greater than the surrounding broadband noise in the airfoil noise spectrum. This noise source occurs when the turbulent boundary layers that form on either side of the airfoil separate from the surfaces of the blunt edge forming two regions of shear in the near wake. Interaction of the vorticity in these shear layers creates a large scale wake instability that results in vortex shedding downstream of the edge.

Spatially correlated unsteady pressures in the trailing edge region are induced by the formation of the periodic vortex street and the resultant unsteady lift force causes far-field sound radiation in the form of a dipole source. When the airfoil is compact (the acoustic wavelength is much greater than the airfoil chord), the sound intensity exhibits a velocity dependence to the sixth power. The non-compact case is a much more efficient radiator as the sound intensity is proportional to the velocity raised to the fifth power.

According to Blake [14], the trailing edge is considered sharp and tonal noise will be negligible for a bluntness parameter of $h_{TE}/\delta^* \leq 0.3$ where h_{TE} is the trailing edge thickness and δ^* is the boundary layer displacement thickness. For larger values of the bluntness parameter $h_{TE}/\delta^* > 0.3$, vortex shedding tonal noise is likely to occur. Care must be taken when using the bluntness parameter to predict the presence of a vortex shedding tone, however. In agreement with Blake [14], Brooks and Hodgson [17] measured a tonal noise contribution from a NACA0012 airfoil with rounded trailing edge and $h_{TE}/\delta^* = 0.48 - 0.64$. They observed the tone to broaden and then diminish with decreasing value of the bluntness parameter. Herr and Dobrzynski [62] on the other hand, reported a bluntness tone from a flat plate with a thin truncated edge of $h_{TE}/\delta^* = 0.18$. They also did not observe any tonal broadening effect for edges of varying thickness.

The vortex shedding frequency is dependent on the size of the separated flow region or the wake thickness. The approximate Strouhal number associated with vortex sound from a blunt trailing edge is [14]

$$\frac{2\pi f y_f}{U_s} \simeq 1 \tag{8.12}$$

where y_f is a wake thickness parameter (defined as the distance between upper and lower regions of shear in the near wake or the inflection points in the wake mean velocity profile) and U_s is the effective local velocity or the effective shear-layer velocity related to the base pressure coefficient, C_{pb}, according to $U_s = U_\infty\sqrt{1 - C_{pb}}$. For truncated and asymmetrically bevelled trailing edges, Blake [14] reports $U_s/U_\infty = 1.03 - 1.25$. An overview of common Strouhal numbers for various edge geometries and Reynolds number ranges is also provided in [14]. In relative agreement with the Strouhal number relationship of Eq. 8.12, Shannon and Morris [118] measured the tonal noise of a flat plate asymmetrically 45° bevelled trailing edge to occur at

$$\frac{2\pi f y_f}{U_\infty} = 1.15 \pm 0.025 \tag{8.13}$$

In this case, the value of the normalised wake thickness parameter was $y_f/h_{TE} \approx 0.5$.

For a NACA0012 airfoil with rounded trailing edge, an alternative Strouhal number dependence based on trailing edge thickness has been reported [17]

$$\frac{f h_{TE}}{U_\infty} = 0.1 \tag{8.14}$$

This Strouhal number relationship has also been found to hold for flat plate models with truncated trailing edges and a wide range of bluntness parameters ($h_{TE}/\delta^* = 0.18 - 0.84$) [62, 93].

Theoretical models that can predict the tonal noise produced by a blunt trailing edge are relatively rare. Roger et al. [112] have developed one of the few analytical vortex shedding noise models, in which a reversed Sears' problem was formulated to determine the unsteady lift distribution that occurs when vorticity is shed from the trailing edge. The far-field sound calculated using a modified version of Amiet's [1] model agreed well with the vortex shedding tone produced by a thick flat plate.

8.2.3 *Turbulent Boundary Layer Trailing Edge Noise*

At high Reynolds numbers or when the airfoil surface is artificially tripped, turbulent boundary layers that are attached at the trailing edge generate broadband noise through an aeroacoustic scattering mechanism. Turbulent flow consists of eddies of various sizes and speeds (or scales) and this gives rise to broadband surface pressure fluctuations near to the trailing edge. The sudden impedance change at the trailing edge scatters the unsteady surface pressure creating acoustic waves [2], as illustrated in Fig. 8.1.

Figure 8.7 shows an example of a typical one-third octave band turbulent trailing edge noise spectrum. It is broadband in nature and features a low frequency peak (denoted f_{peak}) where the sound level is maximum. While this peak has been reported in a number of different trailing edge noise measurement campaigns [12, 18, 61, 65] and computational studies [88, 123], some suggest that it may be due to extraneous noise sources such as facility scattering effects [41, 90]. Doolan and Moreau [42] compiled experimental and numerical turbulent trailing edge noise data sets for a NACA0012 airfoil at a wide range of chord based Reynolds numbers ($4.08 \times 10^5 < Re < 4.06 \times 10^6$) and angles of attack ($0 - 22°$). Taking the mean of all noise data at zero angle of incidence, the Strouhal number associated with the low frequency peak was found to be

$$St_{peak} = \frac{f_{peak}\delta^*}{U_\infty} = 0.069 \tag{8.15}$$

where the boundary layer displacement thickness, δ^*, was calculated using the empirical formulations of Brooks et al. [18]. Other studies have also reported a similar trailing edge noise peak Strouhal number of $St_{peak} = 0.06 - 0.08$ [35, 63]. Increasing the airfoil angle of attack slightly reduces the trailing edge noise peak frequency. This is attributed to a change in the eddy length scales in the outer boundary layer on the suction side near to the trailing edge which results in a shift in the acoustic energy to lower frequencies.

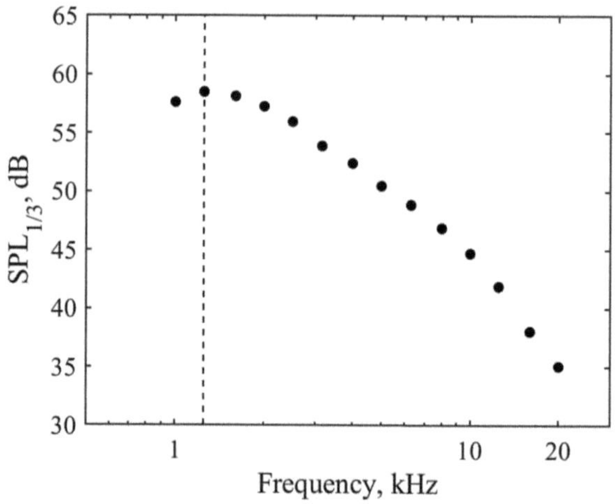

Fig. 8.7 Example one-third octave band turbulent boundary layer trailing edge noise spectrum for a NACA0012 airfoil at $Re = 1.5 \times 10^6$. The vertical dashed line indicates the value of f_{peak}. Adapted from [42]

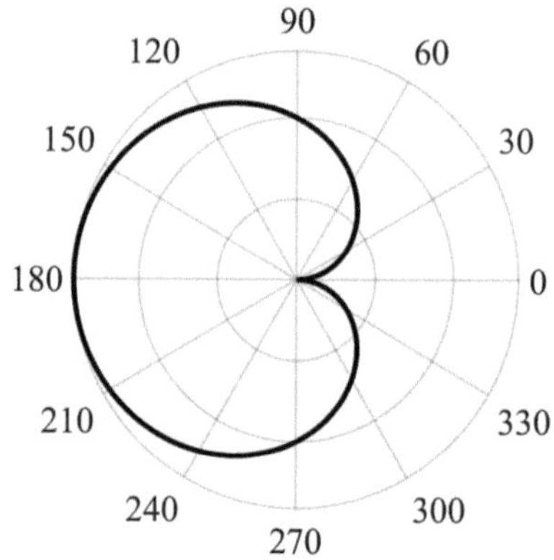

Fig. 8.8 Cardioid trailing edge noise directivity pattern at high frequencies. The origin is at the trailing edge location and the flow direction is from left to right

Turbulent boundary layer trailing edge noise exhibits a dipole-like radiation pattern (similar to that in Fig. 8.2a) when the airfoil is compact and the acoustic wavelength is much larger than the airfoil chord. At high frequencies when the acoustic wavelength is much smaller that the chord, the airfoil approximates a semi-infinite half plane and the trailing edge noise has a cardioid directivity pattern [44]. This directivity pattern is illustrated in Fig. 8.8 and shows the loudest sound is projected upstream while no sound propagates directly downstream. Between these two extremes when the wavelength is on the order of the airfoil chord, the acoustic waves generated at the trailing edge propagate towards the leading edge where they are scattered. This process leads to a directivity pattern close to the original cardioid shape that is modified with additional lobes.

Exact analytical solutions are available for the prediction of trailing edge noise [2, 44, 70] and reviews of the different analytical approaches have been provided by Howe [67] and more recently by Lee et al. [82]. Readers are directed to these

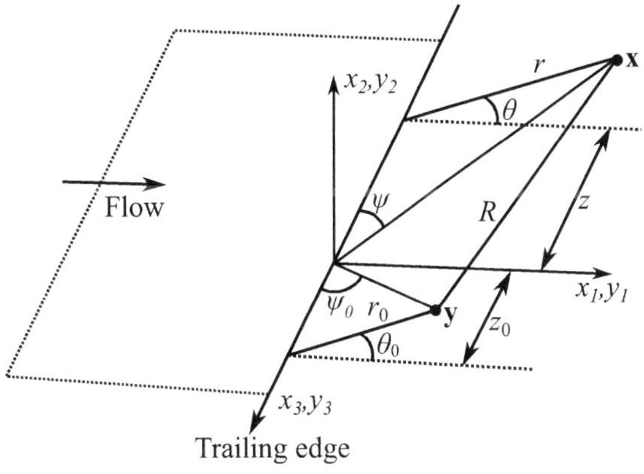

Fig. 8.9 Coordinate system for Ffowcs Williams and Hall's [44] trailing edge noise theory

resources for full derivations; some of the important analytical results are reproduced and discussed here.

Ffowcs Williams and Hall [44] were one of the first to develop an analytical model of the sound generated by edge scattering of low Mach number ($M < 1$) turbulent flow. They considered the case of a rigid, thin, semi-infinite flat plate immersed in a fluid and modelled a turbulent eddy as a quadrupole point source located near the edge. Figure 8.9 shows the coordinate system for Ffowcs Williams and Hall's theory [44] with $\mathbf{x}$ the observer location and $\mathbf{y}$ the position of the turbulent source. Lighthill's [84] equation (Eq. 7.24) which describes aerodynamic noise generation and propagation was used as the basis for the analysis. It was assumed that viscous and nonisentropic conditions can be ignored and turbulent fluctuations (given by the Reynolds stress term $\rho v_i v_j$) are the only source in Lighthill's stress tensor (Eq. 7.25). Taking the Fourier transform, Lighthill's equation can be written as the inhomogeneous Helmholtz equation given by

$$k^2 p^* + \frac{\partial^2 p^*}{\partial y_i^2} = -\left[\frac{\partial^2(\rho v_i v_j)}{\partial y_i \partial y_j}\right]^* \tag{8.16}$$

where k is the acoustic wavenumber and the superscript * denotes the Fourier transform of the physical quantities. The presence of the rigid plate leads to the boundary condition of vanishing normal velocity at the plate surface. The solution to Eq. 8.16 was determined by selecting a Green's function, G, that satisfies this boundary condition and the acoustic field can be calculated by integrating over the volume sources according to

$$4\pi p^*(\mathbf{x}, \omega) = \int (\rho v_i v_j)^* \frac{\partial^2 G}{\partial y_i \partial y_j} dV(\mathbf{y}) \tag{8.17}$$

Ffowcs Williams and Hall [44] then examined noise radiation for the following two cases: (1) turbulent sources within an acoustic wavelength of the edge and (2) turbulent sources many wavelengths from the edge. In both cases, the far-field observer location is assumed to be many wavelengths from both the turbulent flow region and the edge, that is $kr \gg 1$ and $r \gg r_0$, where r is the distance from the edge to the observer location and r_0 is the distance from the eddy centre to the edge.

For turbulence in the near-field, any part of a given eddy must satisfy $2kr_0 \ll 1$. The full solution for acoustic pressure p^* is provided in [44] but it requires estimation of the complex two-point velocity spectrum function. A more useful result is the relation for far-field acoustic intensity, I, given by

$$I \sim \frac{\rho U_\infty^5 \delta^2}{c_0^2 R^2} \tag{8.18}$$

where R is the separation between the observer and source and δ is the characteristic turbulence correlation length scale. Equation 8.18 is an important result that demonstrates the intensity of sound produced at the trailing edge increases in proportion to the velocity raised to the fifth power. This same result was reproduced in later analytical works by Crighton and Leppington [36], Chase [28] and Chandiramani [27].

For an eddy far from the edge, the noise intensity velocity scaling is equivalent to that of a free jet and is given by [44]

$$I \sim \frac{\rho U_\infty^8 \delta^2}{c_0^2 R^2} \tag{8.19}$$

Together, Eqs. 8.18 and 8.19 show that the edge scattering process makes the quadrupole noise sources generated by fluid turbulence in the boundary layer more efficient radiators of sound. A sharp trailing edge is also slightly more efficient than the classical sixth-power scaling predicted in the case of Curle's surface dipoles located well away from the edge.

An alternative class of trailing edge noise prediction models relate the surface pressure fluctuations to acoustic radiation, including those for example by Chandiramani [27], Chase [29], Amiet [2] and Howe [67]. These models are attractive because they do not require computation of the turbulent quadrupole sources in full.

Amiet's [2] formulation utilises the assumption that the boundary layer turbulence is unmodified as it passes over the trailing edge and the coordinate system employed in this model is shown in Fig. 8.10. This assumption allows the noise to be calculated from the wall pressure spectrum imposed by the boundary layer on the airfoil surface without the trailing edge being present. The sound radiation that occurs at the trailing edge is the response of the irrotational flow to the imposition of the Kutta condition and the removal of the non-penetration condition at the trailing edge. This situation is modelled by first imposing a surface pressure on an airfoil that is infinite in both the upstream and downstream directions. Using the theory of Curle [37] (Sect. 7.5), the flow field is represented by a volume quadrupole distribution and a surface dipole distribution. To account for the imaginary downstream airfoil extension, a second solution, obtained from the general Schwartzschild solution, is calculated to cancel

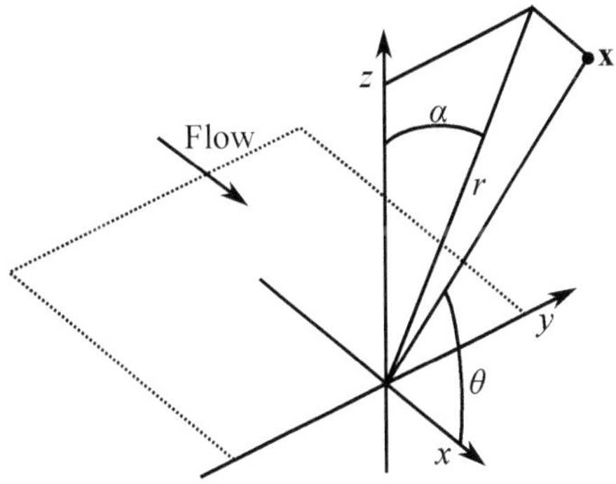

Fig. 8.10 Coordinate system for Amiet's [2] trailing edge noise theory

the dipole distribution in the downstream wake. Following Amiet's formulation [2], the far-field acoustic PSD in the midspan ($y = 0$) plane is given by

$$S_{pp}(\mathbf{x}, \omega) = \left(\frac{\omega c z}{2\pi c_0 \sigma^2}\right)^2 \Lambda_y(\omega) d |L|^2 S_{qq}(\omega) \tag{8.20}$$

where the airfoil has span $2d$ and chord $2c$, the observer is located at $\mathbf{x} = (x, 0, z)$ and σ is the far-field corrected observer location. The term $S_{qq}(\omega)$ is the surface pressure spectrum near the trailing edge, $\Lambda_y(\omega)$ is the spanwise correlation length scale of the surface pressure turbulence and L is the airfoil response function. Equation 8.20 states the sound produced by a boundary layer on one side of the airfoil; for statistically identical boundary layers on both sides, the result should be multiplied by 2.

For turbulent flow past the trailing edge, the airfoil response function is given by [2, 3]

$$|L| = \frac{1}{\Phi}\left[(1+i)\left(\sqrt{\frac{(1+M+K_x/\mu}{1+x/\sigma}}\mathbf{E}^*[2\mu(1+x\sigma)]e^{-i2\Phi} - \mathbf{E}^*[2((1+M)\mu + K_x)]\right) + 1\right]. \tag{8.21}$$

where $\Phi = K_x + \mu(M - x/\sigma)$ and $\mu = M\omega c/U_\infty \beta^2$. It is assumed that the pressure field is convected along the airfoil surface at a constant convection velocity, U_c, so that the axial wavenumber is given by $K_x = \omega/U_c$. Amiet [2] also suggests the following relation for the spanwise correlation length scale

$$\Lambda_y(\omega) = 2.1\frac{U_c}{\omega} \tag{8.22}$$

The wall pressure spectrum, S_{qq}, can be obtained directly from wall pressure measurements or computational modelling. Alternatively, several empirical and semi-empirical models of the wall pressure spectrum underneath a turbulent boundary layer exist [22, 53, 78, 114] and these can be used as input to the trailing edge noise theory. The Goody model [53] (see Sect. 6.8.4.2) is the most favoured for zero pressure gradient turbulent boundary layer flows while Rozenberg et al. [114], Kamruzzaman et al. [78] and Catlett et al. [22] have developed surface pressure models that account for pressure gradient effects at an airfoil trailing edge.

Amiet's formulation in Eq. 8.20 has also been corrected to account for leading edge effects [3, 97, 110]. Roger and Moreau [110] integrated a leading edge back-scattering correction into Amiet's model [2] and found that the leading edge scattering effect can be significant at low frequencies. When the Helmholtz number $kc > 1$, back-scattering effects can be ignored.

The final trailing edge noise prediction method mentioned here is the semi-empirical noise model [18, 105]. These types of models are based on trailing edge noise theory but require additional empirical tuning using experimental or numerical data. They are computationally inexpensive to implement and provide a fast estimate of trailing edge noise spectra. One of the most well known semi-empirical trailing edge noise models is the Brooks, Pope and Marcolini (BPM) model developed by Brooks et al. [18]. The BPM model relates the boundary layer parameters at the trailing edge to noise production. The model formulations were derived from aerodynamic and acoustic data for NACA0012 airfoil models at a wide range of angles of attack and Reynolds numbers ($6.9 \times 10^4 < Re < 1.47 \times 10^6$). While the BPM model provides estimates of several different airfoil self-noise mechanisms (including turbulent boundary layer trailing edge noise, blunt trailing edge vortex shedding noise, laminar-transitional boundary layer tonal noise and tip vortex formation noise), it is most commonly used to predict turbulent boundary layer trailing edge noise. The full BPM model is detailed in a publicly available technical report [18] and it has also been included in a design code used for wind turbine noise prediction named NAFNoise [98]. A general overview of the turbulent boundary layer trailing edge noise model is provided as follows.

The BPM turbulent boundary layer trailing edge noise model is based on the edge scattering formulation of Ffowcs-Williams and Hall [44]. The model accounts for the fact that the noise intensity is proportional to the fifth power of the mean velocity or Mach number, M^5, and the turbulent boundary layer displacement thickness, δ^*, and inversely proportional to the square of the distance between the observer and the trailing edge, r.

The model requires calculation of individual suction and pressure side boundary layer noise contributions, denoted SPL_s and SPL_p, respectively, and an additional noise contribution accounting for the effect of the angle of attack, SPL_α. The angle of attack contribution, SPL_α, also referred to as separated flow noise in the BPM report, is the noise associated with separated boundary layer flow on the suction side of the airfoil at non-zero angle of attack. The general formulations of the three individual noise contributions are

$$SPL_s = 10\log\left(\frac{\delta_s^* M^5 2d\overline{D}_h}{r^2}\right) + A\left(\frac{St_s}{St_1}\right) + (K_1 - 3) + \Delta K_1 \quad (8.23)$$

$$SPL_p = 10\log\left(\frac{\delta_p^* M^5 2d\overline{D}_h}{r^2}\right) + A\left(\frac{St_p}{St_1}\right) + (K_1 - 3) \tag{8.24}$$

$$SPL_\alpha = 10\log\left(\frac{\delta_s^* M^5 2d\overline{D}_h}{r^2}\right) + B\left(\frac{St_p}{St_2}\right) + K_2 \tag{8.25}$$

In Eqs. 8.23–8.25, the subscripts p and s refer to the pressure and suction side of the airfoil, respectively. The parameter $2d$ is the airfoil span and the Strouhal number is calculated as $St = f\delta^*/U_\infty$. Functions A and B define the spectral shape and $\overline{D}_h$ is a directivity function. The model contains a number of tuning parameters including empirical constants K_1, K_2 and ΔK_1 and Strouhal number scaling parameters St_1 and St_2. These functions and tuning parameters depend on the flow configuration (e.g. Reynolds number, angle of attack) and their full definition is too complicated to reproduce here; readers are directed to the BPM report for details.

It follows that the total turbulent boundary layer trailing edge noise spectrum in one-third octave bands is the sum of the 3 individual contributions according to

$$SPL_{TOT} = 10\log\left(10^{SPL_s/10}\right) + 10\log\left(10^{SPL_p/10}\right) + 10\log\left(10^{SPL_\alpha/10}\right) \tag{8.26}$$

Aside from flow configuration (Reynolds number, angle of attack) and geometrical information (observer position), the major inputs into the BPM model are the boundary layer displacement thicknesses, δ_s^* and δ_p^*. These values can be obtained from airfoil boundary layer measurements or an airfoil flow solver such as XFOIL [43]. Alternatively, the BPM model also provides empirical formulations for the boundary layer displacement thickness.

The BPM model is most accurate for the range of flow configurations used to produce the model. Predictions at high Reynolds numbers ($Re > 1.47 \times 10^6$) rely on extrapolated equations. To accurately predict the noise from high Reynolds number applications, further work is needed to extend the BPM model. Several more recent semi-empirical prediction methods have also been developed that can incorporate simulations of the flow physics. Examples include the TNO-Blake model [105] which utilises simulations of the boundary layer flow parameters and a wind turbine noise prediction tool [102].

8.3 Airfoil Noise Control

Having discussed the major airfoil noise generation mechanisms, our attention now turns to controlling airfoil noise. Strategies to reduce airfoil noise can be categorised into active and passive techniques. Examples of active control methods include boundary layer suction or injection and plasma actuation. Active methods are typically less attractive than passive methods due to the complexity of their operation and the additional power requirements or auxiliary equipment needed to manipulate the flow. Passive techniques reduce noise generation through geometrical changes to different parts of the airfoil. A number of different passive noise control strategies have been proposed to reduce airfoil noise inspired by silent owl flight. Owls possess the ability to suppress aerodynamic noise within the audible range of their prey. Compared to non-silent flying birds, the owl's wing has three unique features thought to be responsible for silent flight: (1) comb-like fibres along the leading edge; (2) a compliant trailing edge fringe and (3) a soft downy coating on the surface of the wing [54]. The different passive noise control devices discussed in this section have explored technical application of owl wing attributes to an airfoil in an effort to reduce noise production.

The discussion of passive noise control studies in this section is by no means exhaustive but meant to provide an overview of the different techniques, give an indication of their noise reduction potential and discuss the underlying noise control mechanisms. Readers are referred to recent review papers [73, 82] for additional information.

8.3.1 Serrations

The passive airfoil noise control strategy that has received the most attention in recent years is serrations. Serrations have been applied to both the leading and trailing edge of an airfoil in an effort to target several different noise mechanisms. A typical trailing edge sawtooth serration geometry is shown in Fig. 8.11. Conventional serrations are sinusoidal or sawtooth in shape and defined by a root-to-tip amplitude of $2h_s$ and wavelength λ_s.

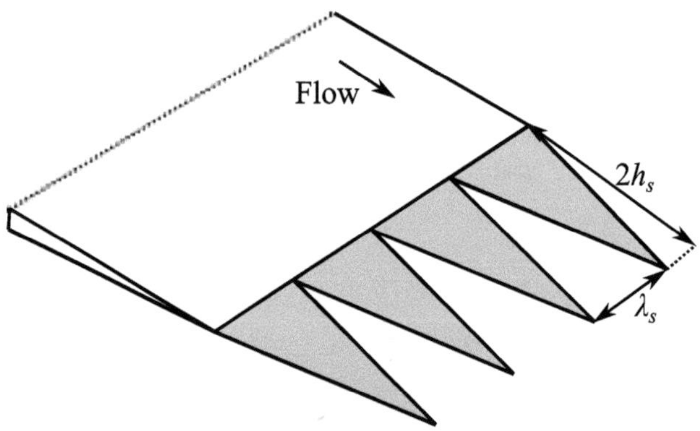

Fig. 8.11 Sawtooth trailing edge serrations. Adapted from [93]

8.3.1.1 Leading Edge Serrations

Serrations applied to the leading edge of an airfoil resemble tubercle (or protuberance or wavy) treatments that are inspired by the undulations present at the leading edge of humpback whale flippers. This morphology is thought to be responsible for a whale's ability to perform aquatic manoeuvrers. When applied to an airfoil, tubercles have been shown to improve aerodynamic performance, especially at high angles of attack near stall [59].

Hersh et al. [66] were one of the first to explore the application of sawtooth serrations to the leading edge of stationary and rotating airfoils with NACA0012 section profile (at Re up to 3.3×10^5) for noise reduction purposes. The serrations were flush mounted to the airfoil surface near the leading edge in a configuration similar to a leading edge trip. The serrations were effective at reducing tonal noise associated with vortex shedding in the wake by between 4 and 8 dB. At high angles of attack (corresponding to stall), broadband noise reductions of 3–5 dB were also recorded.

Shifting focus to the mitigation of leading edge noise, Clair et al. [33] examined the acoustic response of a NACA65 airfoil with sinusoidal leading edge serrations in grid generated isotropic turbulence at a wide range of flow speeds (Re up to 8×10^5). Sound power reductions of approximately 3–4 dB were achieved without introducing an aerodynamic penalty (as indicated by numerical simulations). Using the same model, Narayanan et al. [99] measured a higher maximum attenuation of 7 dB with longer sinusoidal serrations. This indicated that the acoustic performance is impacted by the serration amplitude. A maximum noise reduction of 9 dB was also achieved with the sinusoidal serrations installed at the leading edge of a flat plate. Later, Chaitanya et al. [24] conducted a detailed parametric study to examine the effect of sinusoidal serration amplitude and wavelength on the noise radiation of a flat plate. To achieve the maximum noise reduction, an optimal serration wavelength was identified equivalent to four times the transverse integral length scale of the incoming turbulence. At this wavelength, acoustic sources located at the troughs of adjacent serrations experience destructive interference. The frequency of the noise reduction was found to scale with Strouhal number based on serration amplitude according to $St_{h_s} = fh_s/U_\infty$ and the maximum reduction in sound power level (PWL) at frequency f is $\Delta PWL(f) = 10\log_{10}(fh_s/U_\infty) + 10$. The same acoustic behaviour was also reported when sinusoidal leading edge serrations were applied to real airfoil geometries. Using an analytical approach, Lyu and Azarpeyvand [85] also demonstrated that the serrated edge noise reduction is due to destructive interference effects in the scattered surface pressure induced by the presence of the serrations. They derived a noise radiation model for a serrated sawtooth leading edge based on an iterative form of Amiet's model [1] solved using the Schwarzschild technique. The model predictions were found to be in close agreement with the flat plate measurements of Narayanan et al. [99].

An alternative leading edge serration noise reduction mechanism referred to as the source cut-off effect has also been proposed [79]. In numerical simulations of sinusoidal serrations installed on a flat-plate airfoil, Kim et al. [79] observed that the surface pressure fluctuations in the hill region (located between the serration peak and root) were substantially lower than those at the peak and root locations. This arises from the local sweep angle effect in the hill region (where the sweep angle is significantly higher) and leads to lower radiated sound power levels. Turner and Kim [122] later demonstrated that a secondary horseshoe vortex system is generated by the serrated edge that alters the upstream flow field and enhances the acoustic source strength at the serration root. Thus the root is the dominant noise source location of the serrated leading edge and its acoustic source strength is similar to that of its straight edge counterpart.

To improve acoustic performance, analytical optimisation of the leading edge serration geometry has been explored. Using asymptotic analysis, Lyu et al. [87] showed that to achieve the maximum noise reduction, the serration profile should not contain stationary points; the serration profile must be smooth where possible and feature a sharp, pointed shape (large slope) around any non-smooth points. In accompanying flat plate noise measurements in grid generated turbulence (at Re up to 7.9×10^5), the optimised serrations were found to be superior to conventional sawtooth serrations and achieved an additional 7 dB noise reduction. Slitted and multi-wavelength leading edge serrations have also been explored [23, 25]. Double-wavelength serrations are formed by combining two single-wavelength sinusoidal components of different wavelength, amplitude and phase. The objective of this edge design is to produce two serration roots that are located in close proximity to one another while being separated in the streamwise direction. This leads to destructive interference of the acoustic sources at adjacent root locations and can produce an additional 3 dB far-field noise reduction compared to single-wavelength serrations. Double-wavelength serrations can be adjusted to provide maximum noise reduction at a desired frequency by appropriately selecting the streamwise distance between adjacent root locations.

8.3.1.2 Trailing Edge Serrations

Our attention now turns to the trailing edge of the airfoil. In the discussion that follows, we will focus on the impact of trailing edge serrations on turbulent boundary layer trailing edge noise, unless otherwise stated.

In a series of early landmark analytical studies, Howe [68, 69] derived noise radiation models for a semi-infinite flat plate serrated trailing edge in low Mach number flow. Trailing edges with both sawtooth and sinusoidal serrations were examined with the sawtooth profile found to be more effective in reducing trailing edge noise. According to Howe's theory, trailing edge noise mitigation is attributed to a reduction in the effective spanwise length of the trailing edge that contributes to noise generation. Acoustic radiation occurs efficiently for wavenumber components of the boundary layer pressure field that are normal to the local trailing edge. When a serrated edge is employed, the number of these components decreases as the angle of the trailing edge to the mean flow increases. The magnitude of the noise reduction is dependent on the height and wavelength of the serrations and on the frequency of sound. When the acoustic frequency is high such that $\omega h_s/U_\infty \gg 1$, the theoretical maximum reduction in radiated mean square pressure is proportional to $10\log_{10}[1 + (4h_s/\lambda_s)^2]$ for serrations with a sawtooth profile. The sound generated by large eddies whose length scales are greater than the amplitude of the serrations (low frequency sound) is unaffected by the presence of the serrations, and hence significant noise reductions are only expected in the high frequency region. The largest noise reductions occur when the dimensions of the serrations are of the order of the turbulent boundary layer thickness and when the angle between the mean flow and the local tangent to the wetted surface is less than 45°. This suggests that sharper serrations with a smaller wavelength-to-amplitude ratio λ_s/h_s will result in greater noise reduction.

Experimental work [38, 56, 93] has demonstrated that Howe's [68, 69] models for both sawtooth and sinusoidal serrations tend to overpredict the noise reduction measured in practice and more robust noise models have since been developed [8, 86]. Lyu et al. [86] developed a noise radiation model for a serrated sawtooth trailing edge using a generalised form of Amiet's trailing edge noise theory [2] and application of the Schwarzschild technique. The model suggests that the noise reduction originates from destructive interference effects in the wall pressure field near the serrated trailing edge. While analytical models provide insight into the acoustic radiation effect of a serrated trailing edge, they do not capture their impact on the hydrodynamic field which requires experimental or numerical treatment.

The general trend observed in experimental studies of a serrated trailing edge is that the largest noise reductions, typically on the order of 7 dB in lab scale tests, are achieved at low to mid frequencies and a noise increase is observed at high frequencies. In an early experimental study, Dassen et al. [38] recorded maximum low frequency noise reductions (between 1 and 6 kHz) of 10 and 8 dB for serrated flat plates and NACA airfoils, respectively, at Re up to 1.4×10^6. Oerlemans et al. [103, 104] investigated the reduction of trailing edge noise from model-scale wind turbine blades and the blades of a full-scale 2.3 MW wind turbine by shape optimization and the application of trailing edge serrations. Optimizing the model-scale blade shape for low noise emission and adding trailing edge serrations achieved an overall reduction of 6–7 dB in the radiated noise levels over a variety of flow conditions. Trailing edge serrations applied to a full-scale wind turbine blade were found to decrease the average overall sound levels by up to 3.2 dB at low frequencies below 1 kHz and increase the noise levels above this frequency without any adverse effect on aerodynamic performance. Gruber et al. [55–58] performed one of the largest parametric studies on trailing edge serrations by examining the noise reduction achieved with 36 different sawtooth serration geometries on a NACA 65(12)-10 airfoil at $2.1 \times 10^5 < Re < 8.3 \times 10^5$. Noise reductions of up to 7 dB were achieved at low frequencies, and an increase in noise level was observed at high frequencies. The frequency delimiting a noise reduction and a noise increase corresponds to a constant Strouhal number of $St_\delta = f\delta/U_\infty = 1$. The low frequency noise reduction was attributed to a reduction in the convection velocity and the coherence of the surface pressure field near the edges of the serrations. Meanwhile, the increase in noise at high frequencies was attributed to cross-flow through the roots between adjacent teeth due to the mean pressure difference between the pressure and suction sides of the airfoil. Serrations with small amplitude relative to the trailing edge boundary layer thickness ($h_s/\delta < 0.5$) were found to be inefficient at reducing noise since their height is smaller than the largest eddy size and thus the acoustic scattering efficiency becomes similar to that of a straight trailing edge.

In recent years, focus has turned to examining the complex flow field around individual serrations in an effort to shed light on the noise reduction mechanism [6, 30]. Serrated trailing edge flow is strongly three-dimensional and a pair of counter-rotating streamwise-oriented vortices has been observed to form along the edges of the serrations due to the mean pressure difference between the suction and pressure sides of the airfoil. The convection velocity of these structures normal to the serrated edge is lower than that of the straight edge which results in less efficient radiation of sound. The noise reduction is therefore thought to be associated with a reduction in the scattering efficiency of the serrated edge. However, even at low angles of attack, turbulent flow passes through the empty space between the serrations and this is likely the source of the high frequency noise increase.

In an effort to improve noise reduction potential, several novel serrated trailing edge designs have been explored including slitted [58, 83], randomly shaped [32], poro-serrated [75], slitted-sawtooth, sawtooth-sinusoidal [11] and serrations with combs or slits in the space between them [101]. One noteworthy novel design is the iron-shaped or curved serration [7]. Curved serrations have less free space between the roots of each individual serration compared to conventional designs. This geometrical feature prevents three-dimensional flow motions from occurring near the roots and leads to milder interaction between the flow on the two sides of the airfoil. This in turn leads to a reduction in the scattered pressure at the root of the serrations and an additional 2 dB attenuation in far-field noise compared to conventional serrations.

The capability of trailing edge serrations in reducing airfoil tonal noise in a laminar-transitional flow regime has also been examined [31, 32, 46, 94]. Chong and Joseph [32] applied four different sawtooth serration geometries to the trailing edge of a NACA0012 in this flow regime (at $1 \times 10^5 < Re < 6 \times 10^5$). The serrations were found to inhibit the two-dimensionality of T-S waves at the trailing edge and prevent flow separation from occurring (in cases where a separation bubble was located close to the trailing edge). This disrupts the T-S wave amplification process and leads to a reduction in the amplitude of the tonal noise components. Contrary to the findings of Howe [69], larger reductions in tonal noise were achieved with large serration angles between the mean flow and the local tangent to the wetted surface. A recent LES performed by Gelot and Kim [46] provides additional insight into the tonal noise reduction mechanism. In this study, sawtooth serrations installed at the trailing edge of a Joukowski airfoil at $Re = 2.5 \times 10^5$ achieved 10 dB attenuation in the far-field noise at the primary tone frequency. The tonal noise reduction was attributed to two separate mechanisms, the first being a reduction in the wall pressure difference between the two sides of the airfoil which weakens the scattering at the trailing edge. Secondly, the presence of the serrated edge was found to enhance the phase variation in the wall pressure fluctuations along the span in the vicinity of the edge. This produces destructive interference effects in the wall pressure field and weakens the feedback loop process.

8.3.2 *Porous or Flow Permeable Materials*

An alternative passive airfoil noise control strategy is the use of porous or flow permeable materials. Applying a porous material to an airfoil seeks to replicate the soft plumage featured on an owl wing's upper surface. Porous edge treatments along with fully porous airfoils have been investigated in an effort to minimise leading edge noise and turbulent boundary layer trailing edge noise. As much greater attention has been paid to the application of porous materials in trailing edge noise control, our discussion will begin with this noise mechanism.

8.3.2.1 Porous Materials for Trailing Edge Noise Control

In principle, a porous trailing edge creates a more gradual transition between the solid airfoil surface and the flow in the wake thus weakening the scattering effect of the sharp geometrical discontinuity at the trailing edge. An analytical study by Jaworski and Peake [72] examined the impact of porosity on the Mach number dependency of the far-field trailing edge noise radiated by a semi-infinite flat plate. The porosity was defined by a dimensionless parameter δ_H proportional to $\alpha_H c_0/\omega R_p$, where α_H is the open-area ratio and R_p is the linear dimension of the pores. When the porosity is high ($\delta_H \gg 1$), the acoustic power scattered by a porous plate scales with the sixth power of the flow speed at low frequencies, which is less efficient than the fifth power scaling of a solid edge.

In early experimental tests, entire airfoils were constructed from commercially available porous materials. Geyer et al. [48] constructed airfoils with SD7003 profile from 16 different types of porous materials characterised by their airflow resistivity, r_s. This parameter is a measure of the resistance of the material to the motion of air particles within it. It is defined as the ratio of the pressure drop across a sample of the material, Δp, to the product of the sample thickness, t, and flow velocity, U_∞, according to $r_s = \Delta p / U_\infty t$. At Re up to 8×10^5, the porous airfoils produced more than 10 dB of attenuation in turbulent boundary layer trailing edge noise at low to mid frequencies. At high frequencies, more noise was generated by the porous airfoils compared to their solid counterpart attributed to the increased surface roughness effect. Both the acoustic and aerodynamic performance was found to be dependent on airflow resistivity, although no simple parametric relationships could be defined; in general, increasing the airflow resistivity decreased the noise emission but also produced a decrease in lift and an increase in drag.

To minimise the aerodynamic penalty while maintaining the acoustic advantage, partially porous airfoils and porous inserts at the trailing edge have been investigated. In an extension to their earlier study, Geyer and Sarradj [47] observed that an airfoil with a larger chord-wise porous extent achieves a more significant noise reduction compared to a solid airfoil. However, only a small portion of the trailing edge (the last 5% of the chord) required porous coverage to achieve up to 8 dB of attenuation and no significant lift penalty. Various porous treatments inserted into the trailing edge of a DLR F16 profile (covering 10% of the chord) were studied by Herr et al. [64]. At *Re* up to 1.2×10^6, the porous inserts achieved broadband noise reductions of 2–6 dB at low to mid frequencies (up to 10 kHz depending on the porous material). In agreement with Geyer et al. [48], the level of attenuation was found to be dependent on the airflow resistivity of the porous material rather than its porosity, which was confirmed by covering one side of the porous insert with tape. This implied that a porous trailing edge permits flow communication between the pressure and suction sides of the airfoil in a type of 'pressure release process' which leads to lower edge scattering efficiency. Herr et al. [64] also demonstrated that the pore size is the dominant parameter that determines the contribution of roughness noise at high frequencies. Carpio et al. [21] applied porous metal foam trailing edge treatments that covered 20% of the chord to a NACA0018 airfoil. Compared to a solid airfoil, up to 11 dB of attenuation was achieved at low frequencies accompanied by a high frequency noise increase of up to 10 dB (at $Re = 2.63 \times 10^5$). Rather than focusing on airflow resistivity, the flow permeability of the porous foam was treated as the relevant noise reduction parameter with higher permeability leading to a larger noise reduction. Complimentary numerical simulations performed by Teruna et al. [121] suggested that the permeable metal foam treatments produce a low frequency trailing edge noise reduction for two reasons: by softening the geometrical discontinuity at the trailing edge leading to weaker scattering efficiency and distributing the noise sources over the porous surface in such a way that destructive interference occurs in the scattered pressure field.

As an alternative to using commercially available materials, a porous trailing edge can be produced using additive manufacturing (or 3D printing). This technique provides a high degree of accuracy and ensures homogeneity in the individual pore structures. The simplest way to produce a porous edge with additive manufacturing is to create straight cylindrical channels that directly connect the pressure and suction sides of the airfoil in a rectilinear or grid pattern. This method has been used to produce both fixed wing porous trailing edge treatments [21, 126] and blade extensions with integrated trailing edge porosity for a model-scale wind turbine rig [74]. In an extension to their earlier work [21], Carpio et al. [20] examined the performance of 3D printed porous trailing edge inserts installed on a NACA0018 airfoil. Compared to commercial metal foam, the 3D printed inserts required 3 times higher permeability to achieve similar overall noise reduction. This shows that permeability alone does not determine the level of noise reduction. Applying a large number of different 3D printed porous trailing edge inserts to a NACA 65(12)-10 model, Zhang and Chong [126] examined the impact of various parameters (porosity, pore size and chord-wise extent of the porous treatment) on trailing edge noise production. When the pore size is large, the porosity is small and the porous treatment covers a large portion of the trailing edge, tonal noise may be produced due to bluntness-induced vortex shedding in the wake. A noise reduction of up to 7 dB was achieved (at *Re* up to 6×10^5) with a small pore size (< 1 mm), medium to large porosity and small chord-wise porous coverage. Only 3.7% chord-wise coverage at the trailing edge was necessary to achieve a noise reduction and small pore sizes were also required to reduce the high frequency noise increase.

Additive manufacturing provides great scope to investigate novel and optimised porous edge treatments for trailing edge noise reduction. For example, it has been suggested that gradually increasing the permeability towards the trailing edge may lead to greater noise reduction compared to an edge with constant permeability [113]. Another possibility is creating more complex internal topologies such as the inclusion of an air gap or foam insert at the centre of the porous device which may offer enhanced sound absorption [76]. While these complex edge designs can, in theory, be manufactured with 3D printing, their efficacy is yet to be determined in practice.

8.3.2.2 Porous Materials for Leading Edge Noise Control

Porous materials applied to an airfoil have been shown to produce substantial reductions in leading edge noise of up to 10 dB at selected frequencies [49, 50, 125]. Zamponi et al. [125] examined the use of porous materials in the reduction of airfoil-turbulence interaction noise in a rod-airfoil experiment. This flow configuration consisted of an upstream cylinder whose wake is the turbulent inflow that interacts with the leading edge of a downstream NACA0024 airfoil. The entire airfoil was constructed from melamine foam with a porous exoskeleton and wire mesh covering to protect the internal materials. At *Re* of 3.2×10^5, low frequency noise reduction (below 2 kHz) of up to 4 dB was reported along with a high frequency noise increase up to 12 dB attributed to an increased surface roughness effect. The porous airfoil surface was found to minimise the flow distortion leading to weaker velocity fluctuations at the leading edge and this was suggested to be the mechanism responsible for the noise reduction.

Geyer et al. [50] examined the performance of 3D printed porous leading edges installed on a SD7003 airfoil. The leading edge inserts featured cylindrical channels that directly connected the pressure and suction sides of the airfoil. With only 5% chord-wise coverage at the leading edge, reductions of up to 8 dB were measured at low-to-mid frequencies (between 1 and 4 kHz) at *Re* of 6.5×10^5 in grid generated isotropic turbulence. Larger pores produced a larger noise reduction but this was accompanied by additional high frequency noise of up to 5 dB. The low frequency noise reduction was attributed to three separate mechanisms: (1) dissipation of the turbulent kinetic energy at the porous edge, (2) an increase in the boundary layer thickness at the leading edge which can be thought of as an increase in the airfoil thickness which leads to lower noise production and (3) a shift in the turbulence away from the airfoil surface which attenuates the unsteady surface pressure field at the leading edge.

While earlier studies attributed the noise reduction achieved with porous leading edge treatments to a dissipative mechanism [50, 125], an alternative mechanism was proposed by Chaitanya et al. [26]. They examined porosity applied to a flat plate over its entire surface, the leading edge only and a region downstream of the leading edge. The frequency of the noise reduction was found to scale with the length of the porous section, l, according to fl/U_∞. Employing a porous leading edge resulted in narrow bands of noise attenuation at $fl/U_\infty \approx n - 1/2$, where n is an integer. This suggested that destructive interference was occurring between sound generated at the leading edge and the edge discontinuity located between the porous section and the non-permeable downstream region. They also showed that when pores are used to permit flow communication between the two sides of an airfoil, the unsteady pressure difference convects at a lower velocity compared to that of a rigid airfoil which leads to a reduction in noise radiation efficiency.

A promising new type of porous treatment for the reduction of leading edge noise is chordwise and spanwise varying porosity [9, 10]. The latter attempts to combine the concepts of serrations and porosity while the former design eliminates the discontinuity between the porous leading edge region and the non-permeable downstream region of the airfoil. While the development of gradient porosity is still in its infancy, it has been shown to improve noise reduction compared to edges with evenly distributed porosity possibly by an additional 1–2 dB [10].

8.3.3 Trailing Edge Brushes

Trailing edge brushes have been developed in an effort to emulate the owl wing fringe structure [45, 62]. Similar to porosity, brushes attempt to mitigate the sharp geometrical discontinuity at the trailing edge responsible for scattering turbulent boundary layer sources into far-field noise. Elastic brush extensions installed at the trailing edge of a flat plate have been found to reduce turbulent boundary layer trailing edge noise and bluntness noise contributions by up to 14 dB over a wide frequency range (1 to 16 kHz or $0.02 < St_{\delta *} < 0.2$) at Re up to 7.9×10^6 [62]. Longer elastic brush designs (at least equal in length to the boundary layer thickness) are more effective at reducing trailing edge noise than shorter ones and this is attributed to the decreased permeability of the shorter brush fibres. Brushes have also been effective at reducing the trailing edge produced by realistic airfoil geometries [45, 61]. Finez et al. [45] installed flexible brush extensions at the trailing edge of a NACA65(12)-10 airfoil and achieved 3 dB reduction in turbulent boundary trailing edge noise at frequencies up to 2 kHz at Re to 3.47×10^5. The noise reduction was attributed to a reduction in the spanwise correlation length of turbulent eddies at the trailing edge. It was proposed that the ratio of spanwise velocity correlation length scale to fibre diameter, d_f is $\Lambda_y/d_f \gg 1$ for effective brush design.

8.3.4 Finlets

A recent development in the area of turbulent boundary layer trailing edge noise control is finlet surface treatments developed by Clark et al. [34]. These devices aim to alter the boundary layer turbulence that interacts with the trailing edge rather than reduce the edge scattering efficiency. In the study of Clark et al. [34], 20 different finlet designs were applied to the trailing edge region of a DU96-W180 airfoil (starting at 87% chord). The finlets had variations in thickness, density and extension length past the trailing edge. Their height also varied between 0.1δ and δ. At $Re = 3 \times 10^6$, the finlets achieved up to 10 dB of attenuation in far-field turbulent boundary layer trailing edge noise over a wide frequency range. The finlets produced no lift penalty and only a modest drag increase consistent with the increase in wetted surface area due to the addition of the finlets.

While the fundamental finlet noise reduction mechanism remains unknown, recent studies on parallel arrays of streamwise rods (also termed canopies) placed within a turbulent boundary layer over a flat plate provide some insight [52, 60]. The canopies are suspended above the plate and represent idealised finlets without leading edge support structures. Attenuation is observed to occur in the wall pressure spectra beneath the canopies in three different frequency ranges. At low frequencies, the attenuation frequency scales with the canopy height, h_c, according to fh_c/U_e where U_e is the velocity at the edge of the boundary layer. The low frequency noise reduction is attributed to the formation of an additional shear layer that lifts the large structures in the boundary layer away from the wall. The level of attenuation is reduced at mid frequencies due to additional turbulence produced by the canopy. At high frequencies, the attenuation frequency follows conventional dissipative scaling of $f\nu/U_h^2$ where U_h is the velocity at the canopy leading edge. This suggests that the attenuation in the high frequency range is linked with enhanced dissipation in the boundary layer flow. Numerical simulations of finlet fences applied to a NACA0012 airfoil have also shown that finlets shift the turbulent kinetic energy away from the airfoil surface and towards the top of the finlets [16]. This leads to an increase in the separation distance between the turbulent sources and the trailing edge which may be responsible for the high frequency noise reduction [15].

8.4 Summary

This chapter provides an overview of the major mechanisms of airfoil noise production. Leading edge noise generation is described followed by a discussion of two distinct types of trailing edge noise. The first is tonal trailing edge noise produced by either an aeroacoustic feedback loop or vortex shedding and the second is broadband turbulent boundary layer trailing edge noise. As part of this chapter, several analytical leading and trailing edge noise models have been presented.

The chapter concludes with a discussion of passive devices for airfoil noise control namely serrations, porosity, brushes and finlets. These devices have the potential to reduce airfoil noise in a variety of applications across the energy, aerospace, and transportation sectors. Any noise reduction obtained with these devices may however, come at the cost of some aerodynamic penalty and this will be an important consideration in many industrial situations. While some devices (serrations and porosity) are more well studied than others (brushes and finlets), further research is still needed to gain an improved physical understanding of how different devices affect the local flow at the edge and their role in noise production. This knowledge will help us to optimise the device design given any aerodynamic or acoustic constraints.

References

1. Amiet R (1975) Acoustic radiation from an airfoil in a turbulent stream. J Sound Vib 41:407–420
2. Amiet R (1976) Noise due to turbulent flow past a trailing edge. J Sound Vib 47:387–393
3. Amiet R (1978) Effect of incident surface pressure field on noise due to turbulent flow past a trailing edge. J Sound Vib 57:305–306
4. Arbey H, Bataille J (1983) Noise generated by airfoil profiles placed in a uniform laminar flow. J Fluid Mech 134:33–47
5. Arcondoulis E, Doolan C, Zander A, Brooks L, Liu Y (2019) An investigation of airfoil dual acoustic feedback mechanisms at low-to-moderate Reynolds number. J Sound Vib 460:114887
6. Avallone F, Pröbsting S, Ragni D (2016) Three-dimensional flow field over a trailing edge serration and implications on broadband noise. Phys Fluids 28:117101
7. Avallone F, van der Velden W, Ragni D (2017) Benefits of curved serrations on broadband trailing-edge noise reduction. J Sound Vib 400:167–177
8. Ayton L (2018) Analytic solution for aerodynamic noise generated by plates with spanwise-varying trailing edges. J Fluid Mech 849:448–466
9. Ayton L, Colbrook M, Geyer T, Chaitanya P, Sarradj E (2021) Reducing aerofoil-turbulence interaction noise through chordwise-varying porosity. J Fluid Mech 906:A1
10. Ayton L, Karapiperis O, Awasthi M, Moreau D, Doolan C (2021) Spanwise varying porosity for the enhancement of leading-edge noise reduction. In: AIAA aviation 2021 Forum, AIAA 2021-2191
11. Azarpeyvand M, Gruber M, Joseph P (2013) An analytical investigation of trailing edge noise reduction using novel serrations. In: 19th AIAA/CEAS aeroacoustic conference, AIAA 2013-2009
12. Bahr C, Li J, Cattafesta L (2011) Aeroacoustic measurements in open-jet wind tunnels-an evaluation of methods applied to trailing edge noise. In: 17th AIAA/CEAS aeroacoustics conference, AIAA 2011-2771
13. Bertagnolio F, Madsen H, Bak C, Troldborg N, Fischer A (2015) Aerodynamic noise characterization of a full-scale wind turbine through high frequency surface pressure measurements. Int J Aeroacoust 14:729–766
14. Blake WK (2017) Mechanics of flow-induced sound and vibration, Volume 2: Complex flow-structure interactions. Academic
15. Bodling A, Sharma A (2018) Numerical investigation of low-noise airfoils inspired by the down coat of owls. Bioinspiration & Biomimetics 14:016013
16. Bodling A, Sharma A (2019) Numerical investigation of noise reduction mechanisms in a bio-inspired airfoil. J Sound Vib 453:314–327

17. Brooks T, Hodgson T (1981) Trailing edge noise prediction from measured surface pressures. J Sound Vib 78:69–117
18. Brooks T, Pope D, Marcolini M (1989) Airfoil self-noise and prediction. Technical Report Reference Publication 1218, NASA
19. Buck S, Oerlemans S, Palo S (2018) Experimental validation of a wind turbine turbulent inflow noise prediction code. AIAA J 56:1495–1506
20. Carpio A, Avallone F, Ragini D, Snellen M, van der Zwagg S (2019) 3d-printed perforated trailing edges for broadband noise abatement. In: 25th AIAA/CEAS aeroacoustic conference, AIAA 2019-2458
21. Carpio A, Martinez R, Avallone F, Ragini D, Snellen M, van der Zwagg S (2019) Experimental characterization of the turbulent boundary layer over a porous trailing edge for noise abatement. J Sound Vib 443:537–558
22. Catlett M, Anderson J, Forest J, Stewart D (2016) Empirical modeling of pressure spectra in adverse pressure gradient turbulent boundary layers. AIAA J 54:569–587
23. Chaitanya P, Joseph P (2018) Slitted leading edge profiles for the reduction of turbulence-aerofoil interaction noise. J Acoust Soc Am 143:3494–3504
24. Chaitanya P, Joseph P, Narayanana S, Vanderwel C, Turner J, Kim J, Ganapathisubramani B (2017) Performance and mechanism of sinusoidal leading edge serrations for the reduction of turbulence-aerofoil interaction noise. J Fluid Mech 818:435–464
25. Chaitanya P, Joseph P, Narayanan S, Kim J (2018) Aerofoil broadband noise reductions through double-wavelength leading-edge serrations: a new control concept. J Fluid Mech 855:131–151
26. Chaitanya P, Joseph P, Chong TP, Priddin M, Ayton L (2020) On the noise reduction mechanisms of porous aerofoil leading edges. J Sound Vib 485:115574
27. Chandiramani K (1974) Diffraction of evanescent waves with applications to aerodynamically scattered sound and radiation from unbaffled plates. J Acoust Soc Am 55:19–29
28. Chase D (1972) Sound radiated by turbulent flow off a rigid half-plane as obtained from a wavevector spectrum of hydrodynamic pressure. J Acoust Soc Am 52:1011–1023
29. Chase D (1975) Noise radiated from an edge in turbulent flow. AIAA J 13:1041–1047
30. Chong T, Vathylakis A (2015) On the aeroacoustic and flow structures developed on a flat plate with a serrated sawtooth trailing edge. J Sound Vib 354:65–90
31. Chong T, Joseph P, Gruber M (2010) An experimental study of airfoil instability noise with trailing edge serrations. In: 16th AIAA/CEAS aeroacoustic conference, AIAA 2010-3723
32. Chong T, Vathylakis A, Joseph P, Gruber M (2013) Self-noise produced by an airfoil with non-flat plate trailing-edge serrations. AIAA J 51:2665–2667
33. Clair V, Polacsek C, Le Garrec T, Reboul G, Gruber M, Joseph P (2013) Experimental and numerical investigation of turbulence-airfoil noise reduction using wavy edges. AIAA J 51:2695–2713
34. Clark I, Alexander W, Devenport W, Glegg S, Jaworski J, Daly C, Peake N (2017) Bio-inspired trailing edge noise control. AIAA J 55:740–754
35. Colonius T, Lele S (2004) Computational aeroacoustics: progress on nonlinear problems of sound generation. Prog Aerosp Sci 40:345–416
36. Crighton D, Leppington F (1970) Scattering of aerodynamic noise by a semi-infinite compliant plate. J Fluid Mech 43:721–736
37. Curle N (1955) The influence of solid boundaries upon aerodynamic sound. Proc R Soc Lond A 231(1187):505–514
38. Dassen T, Parchen R, Bruggeman J, Hagg F (1996) Results of a wind tunnel study on the reduction of airfoil self-noise by the application of serrated blade trailing edges. In: Proceedings of the European Union wind energy conference and exhibition, pp 800–803
39. Desquesnes G, Terracol M, Sagut P (2007) Numerical investigation of the tone noise mechanism over laminar airfoils. J Fluid Mech 591:155–182
40. Devenport W, Staubs J, Glegg S (2010) Sound radiation from real airfoils in turbulence. J Sound Vib 329:3470–3483

41. Devenport W, Burdisso R, Borgoltz A, Ravetta P, Barone M, Brown K, Morton M (2013) The Kevlar-walled anechoic wind tunnel. J Sound Vib 332:3971–3991
42. Doolan C, Moreau D (2015) A review of airfoil trailing edge noise with some implications for wind turbines. Int J Aeroacoust 14:811–832
43. Drela M (1989) Xfoil: an analysis and design system for low Reynolds number airfoils. In: Mueller T (ed) Low Reynolds number aerodynamics, Lecture notes in engineering. Springer, Berlin, chap 54, pp 1–12
44. Ffowcs-Williams J, Hall L (1970) Aerodynamic sound generation by turbulent flow in the vicinity of a scattering half plane. J Fluid Mech 40:657–670
45. Finez A, Jondeau E, Roger M, MC J (2010) Broadband noise reduction with trailing edge brushes. In: 16th AIAA/CEAS aeroacoustic conference, AIAA 2010-3980
46. Gelot M, Kim J (2020) Effect of serrated trailing edges on aerofoil tonal noise. J Fluid Mech 904:A30
47. Geyer T, Sarradj E (2014) Trailing edge noise of partially porous airfoils. In: 20th AIAA/CEAS aeroacoustic conference, AIAA 2014-3039
48. Geyer T, Sarradj E, Fritzsche C (2010) Measurement of the noise generation at the trailing edge of porous airfoils. Exp Fluids 48:291–308
49. Geyer T, Sarradj E, Giesler J, Hobracht M (2011) Experimental assessment of the noise generated at the leading edge of porous airfoils using microphone array techniques. In: 17th AIAA/CEAS aeroacoustic conference, AIAA 2011-2713
50. Geyer T, Andreas L, Schrödter M, Schneider M, Sarradj E (2019) Reduction of turbulence interaction noise through airfoils with perforated leading edges. Acta Acust Acust 105:109–122
51. Glegg S, Devenport W (2010) Panel methods for airfoils in turbulent flow. J Sound Vib 329:3709–3720
52. Gonzalez A, Glegg S, Hari M, Ottman M, Deveport W (2019) Fundamental studies of the mechanisms of pressure shielding. In: 19th AIAA/CEAS aeroacoustic conference, AIAA 2019-2403
53. Goody M (2004) Empirical spectral model of surface pressure fluctuations. AIAA J 42(9):1788–1794. https://doi.org/10.2514/1.9433
54. Graham R (1934) The silent flight of owls. Aeronaut J 38:837–843
55. Gruber M (2001) Airfoil noise reduction by edge treatments. Phd thesis, University of Southampton
56. Gruber M, Joseph P, Chong T (2010) Experimental investigation of airfoil self noise and turbulent wake reduction by the use of trailing edge serrations. In: 16th AIAA/CEAS aeroacoustics conference, AIAA 2010-3803
57. Gruber M, Joseph P, Chong T (2011) On the mechanisms of serrated airfoil trailing edge noise reduction. In: 17th AIAA/CEAS aeroacoustics conference, AIAA 2011-2781
58. Gruber M, Joseph P, Azarpeyvand M (2013) An experimental investigation of novel trailing edge geometries on airfoil trailing edge noise reduction. In: 19th AIAA/CEAS aeroacoustics conference, AIAA 2013-2011
59. Hansen K, Kelso R, Dally B (2011) Performance variations for leading-edge tubercles for distinct airfoil profiles. AIAA J 49:185–194
60. Hari M, Szoke M, Deveport W, Glegg S (2021) Understanding pressure shielding by canopies. In: AIAA Scitech 2021 Forum, AIAA 2021-0817
61. Herr M (2007) Design criteria for low-noise trailing-edges. In: 13th AIAA/CEAS aeroacoustics conference, AIAA 2007-3470
62. Herr M, Dobrzynski W (2005) Experimental investigations in low-noise trailing-edge design. AIAA J 43:1167–1175
63. Herr M, Appel C, Dierke J, Ewert R (2010) Trailing-edge noise data quality assessment for CAA validation. In: 16th AIAA/CEAS aeroacoustics conference, AIAA 2010-3877
64. Herr M, Rossignol KS, Delfs J, Mößner M (2014) Specification of porous materials for low-noise trailing-edge applications. In: 20th AIAA/CEAS aeroacoustic conference, AIAA 2014-3041

65. Herrig A, Würz W, Krämer E, Wagner S (2008) New cpv-results of naca 0012 trailing-edge noise. In: International conference on methods of aerophysical research ICMAR, vol 30
66. Hersh A, Soderman P, Hayden R (1974) Investigation of acoustic effects of leading edge serrations on airfoils. J Aircr 11:197–202
67. Howe M (1978) A review of the theory of trailing edge noise. J Sound Vib 61:437–465
68. Howe M (1991) Aerodynamic noise of a serrated trailing edge. J Fluids Struct 5:33–45
69. Howe M (1991) Noise produced by a sawtooth trailing edge. J Acoust Soc Am 90:482–487
70. Howe M (1999) Trailing edge noise at low match numbers. J Sound Vib 225:211–238
71. Hunt J (1973) A theory of turbulent flow round two-dimensional bluff bodies. J Fluid Mech 61:625–706
72. Jaworski J, Peake N (2013) Aerodynamic noise from a poroelastic edge with implications for the silent flight of owls. J Fluid Mech 723:456–479
73. Jaworski J, Peake N (2020) Aeroacoustics of silent owl flight. Ann Rev Fluid Mech 52:395–420
74. Jiang C, Yauwenas Y, Fischer J, Moreau D, Doolan C, Gao J, Jiang W, McKay R, Kingan M (2018) Control of rotor trailing edge noise using porous additively manufactured blades. In: AIAA aviation form, AIAA 2018-3792
75. Jiang C, Fischer J, Moreau D, Doolan C (2019) Experimental investigation of novel porous-serrated treatments on airfoil trailing edge noise reduction. In: 25th AIAA/CEAS aeroacoustic conference, AIAA 2019-2435
76. Jiang C, Moreau D, Fischer J, Doolan C (2021) Additively manufactured sound-absorbing porous structures for airfoil trailing-edge noise control. ASCE J Aerosp Eng 34:04021068
77. Jones L, Sandberg R (2011) Numerical analysis of tonal airfoil self-noise and acoustic feedback-loops. J Sound Vib 330:6137–6152
78. Kamruzzaman M, Bekiropoulos D, Lutz T, Würz W, Krämer E (2015) A semi empirical surface pressure spectrum model for airfoil trailing-edge noise prediction. Int J Aeroacoust 14:833–882
79. Kim J, Haeri S, Joseph P (2016) On the reduction of aerofoil-turbulence interaction noise associated with wavy leading edges. J Fluid Mech 792:526–552
80. Kingan M, Pearse J (2009) Laminar boundary layer instability noise produced by an aerofoil. J Sound Vib 322:808–828
81. Lamb H (1932) Hydrodynamics, 6th edn. Dover Publications
82. Lee S, Ayton L, Bertagnolio F, Moreau S, Chong T, Joseph P (2021) Turbulent boundary layer trailing-edge noise: theory, computation, experiment, and application. Prog Aerosp Sci 126:100737
83. Leon A, Avallone F, Pröbsting S, Ragni D (2016) Piv investigation of the flow past solid and slitted sawtooth serrated trailing edges. In: 54th AIAA aerospaces science meeting AIAA Scitech, AIAA 2016-1014
84. Lighthill MJ (1952) On sound generated aerodynamically I. General theory. Proc Roy Soc Lond Ser A Math Phys Sci 211(1107):564–587. https://doi.org/10.1098/rspa.1952.0060
85. Lyu B, Azarpeyvand M (2017) On the noise prediction for serrated leading edges. J Fluid Mech 826:205–234
86. Lyu B, Azarpeyvand M, Sinayoko S (2016) Prediction of noise from serrated trailing edges. J Fluid Mech 793:556–588
87. Lyu B, Ayton L, Chaitanya P (2019) On the acoustic optimality of lead-edge serration profiles. J Sound Vib 462:114923
88. Marsden O, Bogey C, Bailly C (2008) Direct noise computation of the turbulent flow around a zero-incidence airfoil. AIAA J 46:874–883
89. McAlpine A, Nash E, Lowson M (1999) Boundary-layer instability noise on aerofoils. J Sound Vib 222:753–779
90. Migliore P, Oerlemans S (2004) Wind tunnel aeroacoustic tests of six airfoils for use on small wind turbines. J SolEnergy Eng 126:974–985
91. Mish P, Devenport W (2006) An experimental investigation of unsteady surface pressure on an airfoil in turbulence - part 1: Effects of mean loading. J Sound Vib 296:417–446

92. Mish P, Devenport W (2006) An experimental investigation of unsteady surface pressure on an airfoil in turbulence - part 2: Sources and prediction of mean loading effects. J Sound Vib 296:447–460
93. Moreau D, Doolan C (2013) Noise-reduction mechanism of a flat-plate serrated trailing edge. AIAA J 51:2513–2522
94. Moreau D, Doolan C (2016) The generation of tonal noise from sawtooth trailing-edge serrations at low Reynolds numbers. Aeronaut J 120:971–983
95. Moreau S, Roger M (2005) Effect of angle of attack and airfoil shape on turbulence-interaction noise. In: 11th AIAA/CEAS aeroacoustics conference, AIAA 2005–2973
96. Moreau S, Roger M (2007) Competing broadband noise mechanisms in low-speed axial fans. AIAA J 45:48–56
97. Moreau S, Roger M (2009) Back-scattering correction and further extensions of amiet's trailing-edge noise model. part ii: Application. J Sound Vib 323:397–425
98. Moriarty P (2003) Nafnoise: a program for calculating 2d airfoil noise. Design code, National Wind Technology Center, National Renewable Energy Laboratory
99. Narayanana S, Chaitanya P, Haeric S, Joseph P, Kim J, Polacsek C (2015) Airfoil noise reductions through leading edge serrations. Phys Fluids 27:025109
100. Nash E, Lowson M, McAlpine A (1999) Boundary-layer instability noise on aerofoils. J Fluid Mech 382:27–61
101. Oerlemans S (2016) Reduction of wind turbine noise using blade trailing edge devices. In: 22nd AIAA/CEAS aeroacoustic conference, AIAA 2016-3018
102. Oerlemans S, Schepers J (2009) Prediction of wind turbine noise and validation against experiment. Int J Aeroacoust 8:555–584
103. Oerlemans S, Schepers J, Guidati G, Wagner S (2001) Experimental demonstration of wind turbine noise reduction through optimized airfoil shape and trailing-edge serrations. Technical Report NLR-TP-2001-324, National Aerospace Laboratory NLR
104. Oerlemans S, Fisher M, Maeder T, Kogler K (2009) Reduction of wind turbine noise using optimized airfoils and trailing-edge serrations. AIAA J 47:1470–1481
105. Parchen R (1998) Progress report draw: A prediction scheme for trailing-edge noise based on detailed boundary-layer characteristics. TNO Report HAG-RPT-980023, TNO Institute of Applied Physics
106. Paterson R, Amiet R (1976) Acoustic radiation and surface pressure characteristics of an airfoil due to incident turbulence. In: 3rd AIAA aeroacoustics conference, AIAA 76–571
107. Paterson R, Amiet R (1977) Noise and surface pressure response of an airfoil to incident turbulence. J Aircraft 14:729–736
108. Paterson R, Vogt P, Fink M, Munch CL (1973) Vortex noise of isolated airfoils. J Aircr 10:296–302
109. Probsting S, Scarano F, Morris S (2015) Regimes of tonal noise on an airfoil at moderate Reynolds number. J Fluid Mech 780:407–438
110. Roger M, Moreau S (2005) Back-scattering correction and further extensions of amiet's trailing-edge noise model. part i: Theory. J Sound Vib 286:477–506
111. Roger M, Moreau S (2010) Extensions and limitations of analytical airfoil broadband noise models. Int J Acoust 9:273–305
112. Roger M, Moreau S, Guédel A (2006) Vortex-shedding noise and potential-interaction noise modeling by a reversed sears' problem. In: 12th AIAA/CEAS aeroacoustics conference, AIAA 2006-2607
113. Rossian L, Ewert R, Delfs J (2020) Numerical investigation of porous materials for trailing edge noise reduction. Int J Aeroacoust 19:347–364
114. Rozenberg Y, Robert G, Moreau S (2012) Wall-pressure spectral model including the adverse pressure gradient effects. AIAA J 50:2168–2179
115. Santana L, Christophe J, Schram C, Desmet W (2016) A rapid distortion theory modified turbulence spectra for semi-analytical airfoil noise prediction. J Sound Vib 383:349–363
116. Schumacher K, Doolan C, Kelso R (2014) The effect of a cavity on airfoil tones. J Sound Vib 333(7):1913–1931

117. Sears W (1941) Some aspects of non-stationary airfoil theory and its practical application. J Aeronaut Sci 8:104–188
118. Shannon D, Morris S (2006) Experimental investigation of a blunt trailing edge flow field with application to sound generation. Exp Fluids 41:777–788
119. Tam C (1974) Discrete tones of isolated airfoils. J Acoust Soc Am 55:1173–1177
120. Tam C, Ju H (2011) Airfoil tones at moderate Reynolds number: a computational study. In: 17th AIAA/CEAS aeroacoustics conference, AIAA 2011-2711
121. Teruna C, Manegar F, Avallone F, Ragni D, Casalino D, Carolus T (2020) Noise reduction mechanisms of an open-cell metal foam trailing edge. J Fluid Mech 898:A18
122. Turner J, Kim J (2017) Aeroacoustic source mechanisms of a wavy leading edge undergoing vortical disturbances. J Fluid Mech 811:582–611
123. Wolf W, Azevedo J, Lele S (2012) Convective effects and the role of quadrupole sources for aerofoil aeroacoustics. J Fluid Mech 708:502–538
124. Wright S (1976) The acoustic spectrum of axial flow machines. J Sound Vib 45:165–223
125. Zamponi R, Ragini D, Van de Wyer N, Schram C (2019) Experimental investigation of airfoil turbulence-impingement noise reduction using porous treatment. In: 25th AIAA/CEAS aeroacoustic conference, AIAA 2019-2649
126. Zhang M, Chong T (2020) Experimental investigation of the impact of porous parameters on trailing-edge noise. J Sound Vib 489:115694

Chapter 9
Duct Acoustics

Abstract Ducts are found in a wide range of flow noise applications. These include pipe flows, jet engines and marine applications. The fundamentals of acoustic propagation and radiation from sources (particularly dipole sources) are treated in this chapter. The chapter will provide a fundamental grounding in duct acoustics. It will then focus on a particular methodology for understanding the acoustic propagation in circular ducts with an internal dipole source [7]. The chapter includes with a discussion on engineering methods to predict far-field sound from ducts. We focus on very low Mach number flows.

9.1 Fundamentals

The fundamental acoustics of sound propagation within a duct (or waveguide) will be summarised in this section. First, a general method is presented that provides the acoustic pressure for arbitrary sources and surfaces.

The non-homogeneous wave equation for acoustic pressure field $p' = p'(\mathbf{x}, t)$ driven by source $\mathcal{F}(\mathbf{y}, t)$ is (Eq. 7.3),

$$\frac{1}{c^2}\frac{\partial^2 p'}{\partial t^2} - \frac{\partial^2 p'}{\partial x_i^2} = \mathcal{F}(\mathbf{y}, t) \tag{7.3}$$

Taking Fourier transforms we obtain the Helmholtz equation,

$$\frac{\partial^2 \hat{P}(\mathbf{x}, \omega)}{\partial x_i^2} + k_0^2 \hat{P}(\mathbf{x}, \omega) = -\hat{F}(\mathbf{y}, \omega) \tag{9.1}$$

where $\hat{P}(\mathbf{x}, \omega)$ is the Fourier transform of $p'(\mathbf{x}, t)$, $\hat{F}(\mathbf{y}, \omega)$ is the Fourier transform of $\mathcal{F}(\mathbf{y}, t)$ and $k_0 = \omega/c_0$ is the wave number. As per previous chapters, $\mathbf{x}$ is the location of the observer and $\mathbf{y}$ is the location of the source.

C. Doolan and D. Moreau, *Flow Noise*,
https://doi.org/10.1007/978-981-19-2484-2_9

Using a frequency-based Green's function, a solution for an impulsive point source can be written,

$$\frac{\partial^2 G(\mathbf{x}, \mathbf{y}; \omega)}{\partial y_i^2} + k_0^2 G(\mathbf{x}, \mathbf{y}; \omega) = -\delta(\mathbf{x} - \mathbf{y}) \tag{9.2}$$

Using the concept of reciprocity and the divergence theorem [4], the solution to the Helmholtz equation is,

$$\begin{aligned}\hat{P}(\mathbf{x}, \omega) = &\int_S \left\{ G(\mathbf{x}, \mathbf{y}; \omega) \frac{\partial \hat{P}}{\partial y_j}(\mathbf{y}, \omega) - \hat{P}(\mathbf{y}, \omega) \frac{\partial G}{\partial y_j}(\mathbf{x}, \mathbf{y}; \omega) \right\} n_j \, \mathrm{d}S(\mathbf{y}) \\ &- \int_V G(\mathbf{x}, \mathbf{y}; \omega) \hat{F}(\mathbf{y}, \omega) \, \mathrm{d}V(\mathbf{y})\end{aligned} \tag{9.3}$$

Equation 9.3 is used to solve for the acoustic field inside a duct. This is done using a combination of boundary conditions (walls) and the modal expansion method. The use of a modal solution is very powerful. The acoustic field is represented as a sum of eigensolutions or modes that retain their shape as they propagate in the duct. Each of the modes are orthogonal and the sum of modes form the complete solution for the pressure field inside the duct. Each mode can be thought of as a building block that can be put together to obtain the complete solution. The amplitude of each mode depends on how they are excited, that is, the strength, nature and position of the source in the duct. The Green's function in Eq. (9.3) is not the same as the free-field Green's function introduced earlier in this book (Eq. 7.1). Instead, the Green's function incorporates the duct geometry and wall impedance thus is specialised to a particular problem. In some cases analytical models can be used for the duct modes. This chapter will present the duct acoustic modal solutions for rectangular and circular ducts. The solutions presented satisfy Eq. (9.3) and use Green's functions for the particular duct geometries used. Further, because the modes are orthogonal, we can use the separation of variables method to obtain modal solutions.

9.2 Rectangular Ducts

Consider a hard-walled, infinitely-long rectangular duct as shown in Fig. 9.1. The height of the duct is a and the width b, as shown. On the walls, the boundary conditions can be written,

$$\begin{aligned}\frac{\partial \hat{P}}{\partial x_2}(\mathbf{x}, \omega) = 0 \quad \text{at} \quad x_2 = 0, a \\ \frac{\partial \hat{P}}{\partial x_3}(\mathbf{x}, \omega) = 0 \quad \text{at} \quad x_3 = 0, b\end{aligned} \tag{9.4}$$

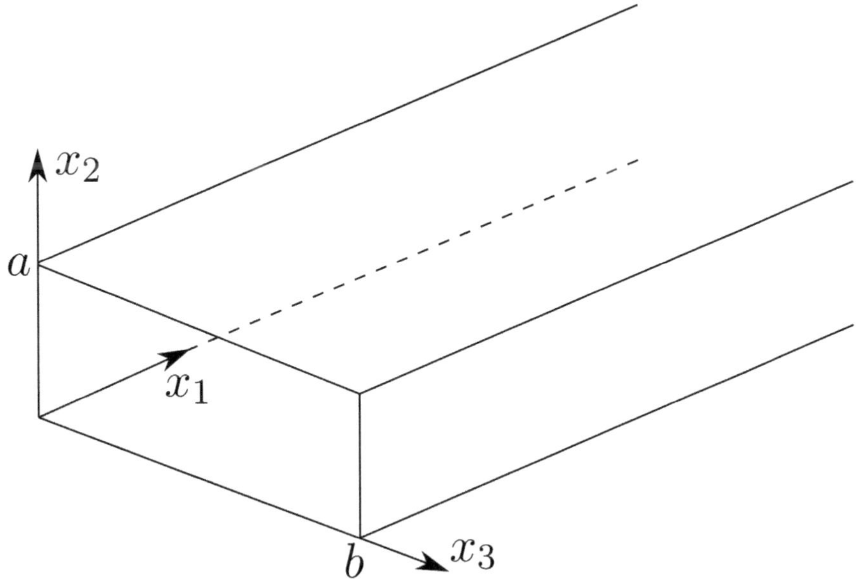

Fig. 9.1 Rectangular duct geometry

As the modes are orthogonal, we use the separation of variables to write the acoustic pressure in terms of three functions [6],

$$\hat{P}(\mathbf{x}, \omega) = \hat{P}(\mathbf{x}) = F(x_1)G(x_2)H(x_3) \tag{9.5}$$

where we drop the ω as it is implied that that each solution is at a fixed frequency. Combining with the boundary conditions (Eq. 9.4), the functions are defined in terms of a axial function (F) and standing wave shapes (G, H) across the duct,

$$F(x_1) = \mathrm{e}^{\mp i\alpha_{mn}x_1} \qquad n = 0, 1, 2, \ldots\ m = 0, 1, 2, \ldots \tag{9.6}$$

$$G(x_2) = \cos(\alpha_n x_2) \qquad n = 0, 1, 2, \ldots \tag{9.7}$$

$$H(x_3) = \cos(\alpha_m, x_3) \qquad m = 0, 1, 2, \ldots \tag{9.8}$$

where the eigenvalues are defined,

$$\alpha_n = \frac{n\pi}{a} \qquad n = 0, 1, 2, \ldots \tag{9.9}$$

$$\alpha_m = \frac{m\pi}{a} \qquad m = 0, 1, 2, \ldots \tag{9.10}$$

and the wavenumber,

$$\alpha_{mn} = \sqrt{\left(\frac{\omega}{c_0}\right)^2 - \alpha_n^2 - \alpha_m^2} \tag{9.11}$$

Therefore the pressure field inside the duct is the sum of right- and left-running (mn) modes,

$$\hat{P} = \sum_{n=0}^{\infty}\sum_{m=0}^{\infty} \cos(\alpha_n x_2)\cos(\alpha_m x_3)(A_{mn}e^{-i\alpha_{mn}x_1} + B_{mn}e^{i\alpha_{mn}x_1}) \qquad (9.12)$$

where the negative in the exponent represents the right running wave and the coefficients A_{mn} and B_{mn} are determined from the source location, strength and type (monopole, dipole, etc.).

The mode where $m = n = 0$ represents the plane wave mode used in one-dimensional acoustics. For higher values of mode numbers, the situation becomes more interesting depending on the value of the wavenumber α_{mn}. If α_{mn} is real, then the function $F(x_1)$ is a complex harmonic and wave propagation occurs in the duct. On the other hand, if α_{mn} is imaginary, then $F(x_1)$ is a decaying exponential function. Here the waves are evanescent and do not propagate.

The cut-off frequency $(\omega_{c,mn})$ is the point at which the (mn)th mode switches from being evanescent to propagating. It can be found by setting $\alpha_{mn} = 0$ and solving Eq. (9.11),

$$\omega_{c,mn} = c_0\sqrt{\alpha_n^2 + \alpha_m^2} \qquad (9.13)$$

The phase velocity for the (mn)th mode is,

$$c_{ph,mn} = \frac{\omega}{\alpha_{mn}} \qquad (9.14)$$

At high frequencies, where the wave number α_{mn} is real, the phase speed approaches the speed of sound c_0. As the frequency approaches the cut-off frequency, the phase speed tend to infinity. Once the frequency is below cut-off, the wave number is complex, the mode is evanescent and there is no propagation (the phase speed becomes complex).

9.3 Circular Ducts

Figure 9.2 illustrates a hard-walled infinitely long circular duct with radius a. Note that we change from the cartesian coordinate system to the cylindrical coordinate system $\mathbf{x} = (r, \theta, x)$.

The Helmholtz equation in cylindrical coordinates is,

$$\frac{\partial^2 \hat{P}(\mathbf{x},\omega)}{\partial x^2} + \frac{\partial^2 \hat{P}(\mathbf{x},\omega)}{\partial r^2} + \frac{1}{r}\frac{\partial \hat{P}(\mathbf{x},\omega)}{\partial r} + k_0^2\hat{P}(\mathbf{x},\omega) = -\hat{F}(\mathbf{y},\omega) \qquad (9.15)$$

We now set $\hat{P}(\mathbf{x},\omega) = \hat{P}$ for convenience.

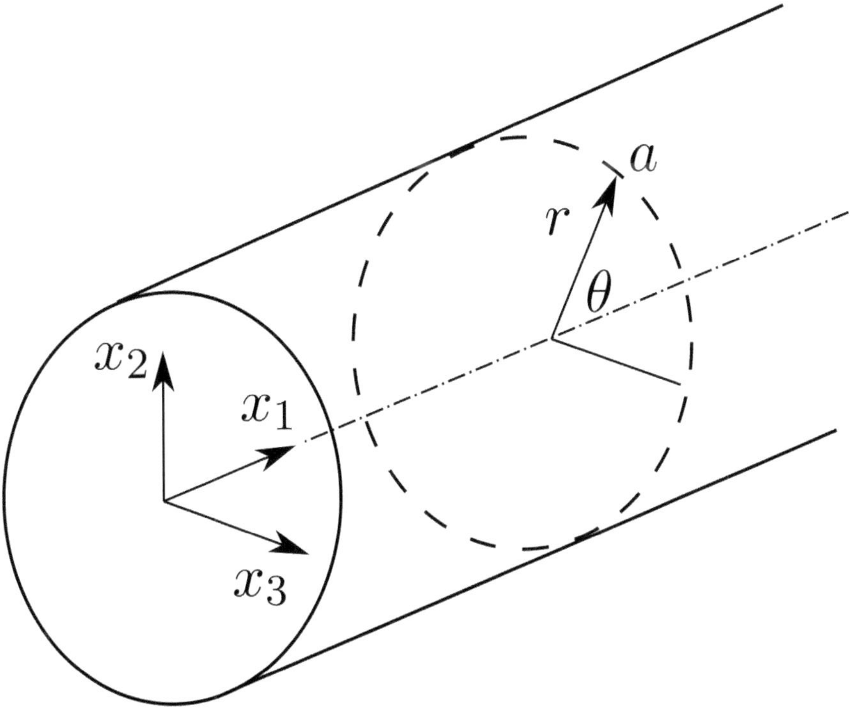

Fig. 9.2 Circular duct geometry

For the hard-walled cylinder, the boundary condition is,

$$\frac{\partial \hat{P}}{\partial r} = 0 \quad \text{at} \quad r = a \tag{9.16}$$

Modal solutions are again found using the separation of variables method,

$$\hat{P}(r, \theta, x) = F(x)G(r)H(\theta) \tag{9.17}$$

where the three terms are independent in space. The circumferential term $H(\theta)$ represents tangential (or spinning) modes,

$$H(\theta) = \mathrm{e}^{-im\theta} \tag{9.18}$$

where $m = 0, \pm 1, \pm 2, \ldots$ is often referred to as the circumferential eigenvalue [6].

The radial term $G(r)$ represent modes across the duct,

$$G(r) = J_m \left(\frac{\mu'_{mn} r}{a} \right) \tag{9.19}$$

Table 9.1 Tabulated values of μ'_{mn}

n m	0	1	2	3
1	3.8317	1.8412	3.0542	4.2012
2	7.0156	5.3314	6.7061	8.0152
3	10.173	8.5363	9.9695	11.346
4	13.3237	11.706	13.170	14.586

where J_m is Bessel function of the first kind of order m (Sect. 3.3). To satisfy the boundary condition (Eq. 9.16), μ'_{mn} is the nth zero of J'_m, so that,

$$J'_m\left(\frac{\mu'_{mn}r}{a}\right) = \frac{\partial J_m\left(\frac{\mu'_{mn}r}{a}\right)}{\partial r} = 0 \quad \text{when} \quad r = a \tag{9.20}$$

where μ'_{mn} is referred to as the radial eigenvalue [6].

The axial mode is,

$$F(x) = \mathrm{e}^{\mp i\alpha_{mn}x} \tag{9.21}$$

where the axial wavenumber (or axial eigenvalue) for the (mn)th mode is

$$\alpha_{mn} = \sqrt{\left(\frac{\omega}{c_0}\right)^2 - \left(\frac{\mu'_{mn}}{a}\right)^2} \tag{9.22}$$

The values of μ'_{mn} are tabulated in Table 9.1. Note there is a special case when $m = n = 0$ that sets $\mu'_{00} = 0$. This is otherwise known as the plane wave case normally associated with one-dimensional duct acoustic wave motion.

The total pressure field is the sum of (mn) modes [1],

$$\hat{P} = \sum_{m=0}^{\infty}\sum_{n=0}^{\infty}(A_{mn}\mathrm{e}^{-i\alpha_{mn}x} + B_{mn}\mathrm{e}^{i\alpha_{mn}x})J_m\left(\frac{\mu'_{mn}r}{a}\right)\mathrm{e}^{-im\theta} \tag{9.23}$$

where the values of A_{mn} and B_{mn} are set by the source type, location and strength.

The phase velocity is,

$$c_{ph,mn} = \frac{\omega}{\alpha_{mn}} \tag{9.24}$$

and the cut-off frequency is,

$$\omega_{c,mn} = \frac{\mu'_{mn}c_0}{a} \tag{9.25}$$

The acoustic field now can be imagined as the linear sum of different orthogonal modes. At low frequencies below the first cut-off frequency, only plane waves propagate in the duct. Higher order propagating modes can only exist (they 'cut-on') when the first cut-off frequency is passed. The first time this occurs in a circular duct is for the $n = 1, m = 1$ mode which has $\mu'_{11} = 1.8412$.

9.4 Sources in a Circular Duct

The Green's function for a monopole source within an infinite, hard-walled cylindrical duct is [5],

$$G(\mathbf{x}, \mathbf{y}; \omega) = G = \frac{e^{ik_0|x-x_0|}}{2i\pi a^2 k_0} \sum_{m=0}^{\infty} \sum_{n=0}^{\infty} \epsilon \frac{J_m\left(\frac{\mu'_{mn} r}{a}\right) J_m\left(\frac{\mu'_{mn} r_0}{a}\right)}{\sigma_{mn}} \cos\left(m(\theta - \theta_0)\right) \times e^{i\alpha_{mn}|x-x_0|} \tag{9.26}$$

where the source is located at $\mathbf{y} = (r_0, \theta_0, x_0)$, $\epsilon = 1/2$ for $m = 0$ and $\epsilon = 1$ otherwise. The scaling term σ_{mn} is,

$$\sigma_{mn} = \pi a^2 \left(1 - \frac{m^2}{\mu'^2_{mn}}\right) J^2_m(\mu'_{mn}) \tag{9.27}$$

Using Eq. (9.1) and the rigid wall boundary conditions, we can write an expression for the pressure generated by a monopole source of strength q_0 as,

$$\hat{P}(\mathbf{x}, \omega) = -\int_V q_0 G(\mathbf{x}, \mathbf{y}; \omega)\, \mathrm{d}V(\mathbf{y}) \tag{9.28}$$

For a dipole source, the Green's function is the spatial derivative of Eq. (9.26) [2]. For axially-oriented dipoles Tomko et al. [7] shows,

$$\hat{P}(\mathbf{x}, \omega) = \int_V |\mathbf{F}_x| \frac{\partial G}{\partial x_0}\, \mathrm{d}V(\mathbf{y}) \tag{9.29}$$

where $|\mathbf{F}_x|$ is the amplitude of an axially-orientated fluctuating force density (the oscillating force amplitude divided by the volume it acts over) at frequency ω.

The axial-derivative of the Green's function is [7],

$$
\begin{aligned}
\frac{\partial G}{\partial x_0} &= (-1)^b \frac{\mathrm{e}^{ik_0|x-x_0|}}{2i\pi a^2} \\
&+ (-1)^b \sum_{m=0}^{\infty}\sum_{n=0}^{\infty} \epsilon \frac{J_m\left(\frac{\mu'_{mn} r}{a}\right) J_m\left(\frac{\mu'_{mn} r_0}{a}\right)}{\sigma_{mn}} \cos\left(m(\theta-\theta_0)\right) \\
&\times \mathrm{e}^{i\alpha_{mn}|x-x_0|}
\end{aligned}
\tag{9.30}
$$

where $b = 0$ if $x \geq x_0$, otherwise $b = 1$.

9.5 Acoustic Radiation from a Duct Exit

Until this point, the analysis has focussed upon infinitely-long ducts. The problem now remains of estimating the acoustic pressure radiated from a duct exit. We now assume a semi-infinite duct with one end having an opening that faces an unbounded acoustic medium. The problem to solve now is to use the pressure at the duct exit to calculate the free-field acoustic pressure. To do this, we use the Kirchoff integral [3],

$$
\hat{P}(\mathbf{x},\omega) = \int_S \left(\hat{P}(\mathbf{y},\omega) \frac{\partial G(\mathbf{x},\mathbf{y};\omega)}{\partial y_i} - G(\mathbf{x},\mathbf{y};\omega) \frac{\hat{P}(\mathbf{y},\omega)}{\partial y_i} \right) n_i \, \mathrm{d}S \tag{9.31}
$$

where $\mathbf{x} = (r_e, \theta_e)$ is the position of the observer in the free-field outside the duct in spherical coordinates, which is assumed to lie on the $x_1 - x_2$ plane. The source position, $\mathbf{y} = (r_0, \phi_0)$ is on the exit plane of the duct (see Fig. 9.3). The Green's function used in Eq. (9.31) can be arbitrary, which can take into account the geometry of the duct exit. Here, we use the free-field Green's function (Eq. 7.1) to obtain a solution in the far-field [3], which is commonly used. Additional complexity can be included with more complicated Green's functions that can include baffles/flanges or other geometries.

Taking a single mode in the duct and integrating over the duct exit, the pressure in the far-field is [3],

$$
\hat{P}(\mathbf{x},\omega) = B_{mn} \int_S \left(\frac{\partial G(\mathbf{x},\mathbf{y};\omega)}{\partial x} - i\alpha_{mn} G(\mathbf{x},\mathbf{y};\omega) \right) J_m\left(\frac{\mu'_{mn} r_0}{a}\right) \mathrm{e}^{-im\phi_0} \, \mathrm{d}S \tag{9.32}
$$

This can be further reduced [3] to yield an expression of the pressure in the far-field for a single mode,

$$
\hat{P}(\mathbf{x},\omega) = -2\pi i B_{mn} J_m(\mu'_{mn}) D_{mn}(\theta_e) \frac{a\mathrm{e}^{ik_0 r_e}}{4\pi r_e} \tag{9.33}
$$

where the directivity of the (mn)th mode is,

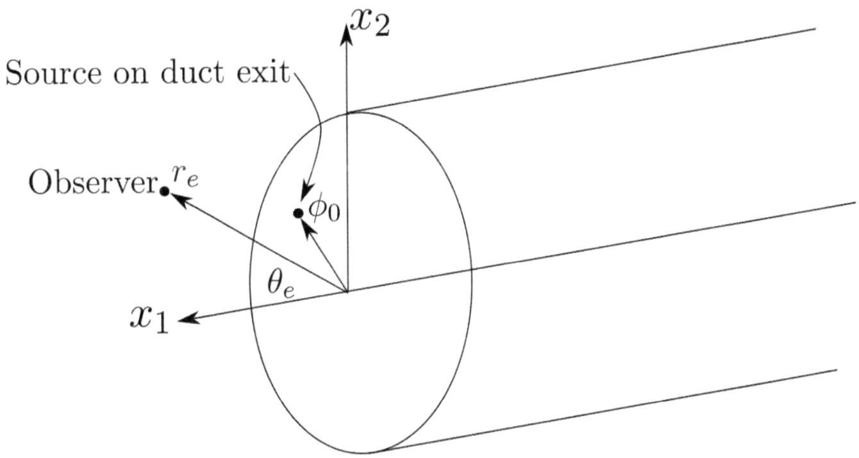

Fig. 9.3 Geometry for acoustic radiation from a duct exit

$$D_{mn}(\theta_e) = \frac{(k_0 \cos\theta_e + \alpha_{mn})(k_0 \sin\theta_e) J'_m(k_0 a \sin\theta_e)}{\mu'^2_{mn} - k_0^2 \sin^2\theta_e} \tag{9.34}$$

Alternatively, numerical solutions of Kirchoff's integral can be used.

9.6 Scaling Model

An engineering model of noise radiation from axially-aligned dipoles within ducts is presented in this section. It is useful for those considering the effects of enclosing a dipole source (such as a propeller) within a duct.

Blake [2] develops some relationships for rectangular ducts that can be assumed to be valid for all types of cross-section. Consider a fluctuating force, aligned with the axis of the duct and placed at the duct centre. For small frequencies below the cut-off frequency (so that $\sqrt{k_0^2 A_D} << 1$, A_D being the area of the duct), only plane waves exist in the duct. In this case, the pressure in the duct is,

$$\hat{P}(\mathbf{x}, \omega) = \frac{1}{2A_D} |\mathbf{F}_x| e^{\pm i k_0 x} \tag{9.35}$$

Hence, the pressure spectral density is,

$$\Phi_{pp}(\mathbf{x}, \omega) = \frac{\Phi_{FF}(\omega)}{4A_D^2} \tag{9.36}$$

where $\Phi_{FF}(\omega)$ is the force spectral density of the source.

We now place the oscillating force in a finite duct, so that the sound within the duct can pass into the free field. The source is placed within the duct so that sound radiates either side of the force. Given that the sound power spectral density is defined,

$$\mathcal{W}(\omega) = \int_S \frac{\Phi_{pp}(\omega)}{\rho_0 c_0} \, \mathrm{d}S \tag{9.37}$$

We can write the total sound power spectral density from one-side of the duct as,

$$\mathcal{W}_D(\omega) = \frac{\Phi_{FF}(\omega)}{4\rho_0 c_0 A_D} \tag{9.38}$$

The acoustic pressure radiated to the free-field from one-side of the duct can be determined using an impedance [2],

$$\hat{P}_a(\mathbf{x}, \omega) = Z_a \hat{U}(L, \omega) \tag{9.39}$$

where Z_a is the impedance at the opening and $\hat{U}(L, \omega)$ is the Fourier transform of the fluid velocity at the duct exit where $x = L$ at the duct exit and the length of the duct is $2L$.

For low frequencies where plane waves only are considered, the impedance can be modelled as piston in a baffled or flanged plate [2],

$$Z_a = \rho_0 c_0 \left[\left(\frac{k_0^2 A_D}{2\pi} \right) + \frac{8i}{3\pi^{3/2}} \sqrt{k_0^2 A_D} \right] \tag{9.40}$$

The sound power density radiated from one-side of the duct is [1],

$$\mathcal{W}_{\text{Flanged}} = \text{Re}(Z_a) \Phi_{uu}(\omega) A_D \tag{9.41}$$

where $\Phi_{uu}(\omega)$ is the velocity spectral density at the duct exit. The velocity at the duct exit can be written in terms of the pressure,

$$\hat{U}(\mathbf{x}, \omega) = \frac{1}{i\omega\rho_0} \frac{\partial \hat{P}}{\partial x} \tag{9.42}$$

Differentiating Eq. (9.35) and converting to the velocity spectral density,

$$\Phi_{uu}(\omega) = \frac{\Phi_{FF}(\omega)}{4\rho_0 c_0 A_D^2} \tag{9.43}$$

Combining with Eq. (9.41), we obtain an expression for the sound power density radiated from one-side of a flanged duct,

$$\mathcal{W}_{\text{Flanged}} = \frac{k_0^2 \Phi_{FF}}{8\pi \rho_0 c_0} \tag{9.44}$$

with the sound power density from both sides simply double this value. Similarly, an unflanged duct is double the sound power again [2].

It is informative to compare the radiation from ducts with equivalent dipoles in a free-field. For an equivalent dipole (as assumed above) but in a free-field rather than a duct, we can write,

$$\mathcal{W}_{\text{Free}}(\omega) = \frac{k_0^2 \Phi_{FF}(\omega)}{12\pi \rho_0 c_0} \tag{9.45}$$

We can now the ratio of sound power for a flanged one-sided duct to that of a free-field,

$$\frac{\mathcal{W}_{\text{Flanged}}}{\mathcal{W}_{\text{Free}}} = \frac{3}{2} \tag{9.46}$$

Similarly, the same ratio for the unflanged one-sided duct is,

$$\frac{\mathcal{W}_{\text{Unflanged}}}{\mathcal{W}_{\text{Free}}} = 3 \tag{9.47}$$

If a finite duct is assumed, the flanged and unflanged ratios become 3 and 6, respectively. These results can be used to guide the design of ducted propellers. For example, we can expect that the radiated sound power to be 3 to 6 times greater (at low frequencies $\sqrt{k_0^2 A_D} << 1$) when a finite duct is placed around the propeller. Thus the designer should be aware that fluctuating loads on the propeller are even more important when a duct is involved. Ducted propellers often contain stators and support struts that can create unsteady loads. Further, the casing boundary layer is also a source of unsteady noise in a ducted propeller. Finally, plane wave reflections from the ends of the duct can create so-called 'organ-pipe' modes. The energy in these modes can cause additional structural excitation and efficient radiation from the duct casing [7].

To conclude this section, we consider high frequencies and source location. When $\sqrt{k_0^2 A_D} >> 1$, the duct outlet impedance approaches $\rho_0 c_0$ and many duct modes are excited; the result is as frequency is increased, the acoustic radiation from the duct approaches that from a the same source in a free-field [2]. Finally, the results presented in this section assume that the source is well away from the duct ends. If the source (propeller) is placed close to the end of the duct, we can expect that the sound radiated by the duct will also include scattering modes (including evanescent waves) from the ends and possibly direct radiation from the source itself if close to the end of the duct. Analysis of these situations call for more advanced analysis than presented here in conjunction with experimental measurements.

9.7 Summary

This chapter provides an overview of duct acoustics. After the fundamentals are presented, solutions for infinitely long rectangular and circular ducts are given. A methodology for incorporating monopole and axially-oriented dipoles within ducts is then described. Acoustic radiation from a semi-infinite duct is then included before the chapter concluding with the provision of a scaling model for acoustic radiation from finite-length ducts.

References

1. Blackstock DT (2001) Fundamentals of physical acoustics. Acoust Soc Am
2. Blake WK (2017) Mechanics of flow-induced sound and vibration, vol 1: General concepts and elementary sources, 2nd edn. Academic press, London
3. Glegg S, Devenport W (2017) Aeroacoustics of low Mach number flows: fundamentals, analysis, and measurement. Academic Press
4. Howe MS (1998) Acoustics of fluid-structure interactions. Cambridge University Press, Cambridge
5. Morse PM, Ingard KU (1986) Theoretical acoustics. Princeton Universty Press, Princeton, NJ, USA
6. Rienstra SW, Hirschberg A (2013) An introduction to acoustics
7. Tomko JR, Stephens DB, Economon T, Morris SC (2019) Experiments and analysis of the internal wall pressure of a ducted rotor. J Sound Vib 451:84–98. https://doi.org/10.1016/j.jsv.2018.12.026

Index

C. Doolan and D. Moreau, *Flow Noise*,
https://doi.org/10.1007/978-981-19-2484-2

MIX
Papier aus verantwortungsvollen Quellen
Paper from responsible sources
FSC® C105338

If you have any concerns about our products,
you can contact us on
ProductSafety@springernature.com

In case Publisher is established outside the EU,
the EU authorized representative is:
Springer Nature Customer Service Center GmbH
Europaplatz 3, 69115 Heidelberg, Germany

Printed by Libri Plureos GmbH
in Hamburg, Germany